Real American Ethics

ALBERT BORGMANN

The University of Chicago Press

Chicago & London

Real American Ethics

Taking Responsibility for Our Country

ALBERT BORGMANN is a Regents Professor of Philosophy at the University of Montana. He is the author of *Technology and the Character of Contemporary Life, Crossing the Postmodern Divide,* and *Holding On to Reality: The Nature of Information at the Turn of the Millennium,* all published by the University of Chicago Press.

The University of Chicago Press, Chicago 60637
The University of Chicago Press, Ltd., London

Printed in the United States of America

15 14 13 12 11 10 09 08 07 06   1 2 3 4 5

ISBN-13: 978-0-226-06634-9 (cloth)
ISBN-10: 0-226-06634-7 (cloth)

Part of chapter 10 was previously published as "Information and Inhabitation," in *Design Philosophy Papers: Collection Two,* ed. Anne-Marie Willis (Ravensbourne, Australia: D/E/S Publications, 2005), pp. 10–19. Part of chapter 10 appeared in "Everyday Fortitude," *The Christian Century,* Nov. 14, 2001, pp. 16–21. Copyright © 2001 Christian Century. Reprinted by permission. Part of chapter 12 was previously published as "A Moral Conception of Commodification," in *The Moralization of the Markets,* ed. Nico Stehr, Christoph Henning, and Bernd Weiler (New Brunswick, NJ: Transaction Publishers, 2006), pp. 193–211.

Library of Congress Cataloging-in-Publication Data

Borgmann, Albert.
Real American ethics : taking responsibility for our country / Albert Borgmann.
p. cm.
Includes bibliographical references (p.) and index.
ISBN-13: 978-0-226-06634-9 (cloth : alk. paper)
ISBN-10: 0-226-06634-7 (cloth : alk. paper)
1. Ethics—United States. 2. United States—Moral conditions. I. Title.

BJ352.B67 2006
170.973—dc22   2006016449

♾ The paper used in this publication meets the minimum requirements of the American National Standard for Information Sciences—Permanence of Paper for Printed Library Materials, ANSI Z39.48-1992.

FOR NANCY

Contents

Preface

This book is my attempt to come to terms with the country I love. I first came here in September of 1958. I had worked my way over as a dish washer on the passenger ship *Arosa Sky*. We steamed into New York harbor. I saw the Statue of Liberty. Someone met me at the dock and took me to the Trailways depot to put me on a bus to Austin, Texas. Did I want a return ticket, valid for a year? I said yes. The man at the window began to write the ticket. I thought about it and said to my guide: No, one way. She said to the ticket writer: He changed his mind. Had I not, I would not have met Nancy; we would not have three daughters now and six grandchildren.

Although the impetus for this book was personal, the execution has been philosophical. What is it to come to philosophical terms with one's country? Philosophers have been educated and are given the time to think through problems that in the lives of other people occur as troubling experiences, irritating puzzles, or flashes of insight. To get to the bottom of these problems and to show how they hang together, some depth and breadth are needed, depth of analysis and breadth of information. From these two dimensions a moral picture has to emerge of the life that we typically lead for better or worse and of the life that we are capable of, but fail to live up to.

While mainstream philosophers in this country have been good at fashioning tools of analysis, they have largely stayed away from drawing a moral portrait of how and what we are actually doing and what we could and should be doing. One reason is that to get your arms around this country, you have to check your philosophical intuitions against what well-positioned observers have seen and especially what social scientists have discovered. This makes for a broad and varied canvas. But it is finally more than a picture and a portrait. It is a story that, although unfinished, has a moral.

INTRODUCTION

The Place of Ethics

CHAPTER ONE

Real American Ethics

The Scope and Coherence of American Ethics

How can anyone hope to set down ethics for so large and populous a country as the United States? Well, you have to remember that the great philosophers of the modern era meant to define ethical principles not just for this continent and our time but for all times and all places. Such universal principles, however, have got to be thin or false. An ethics that matters must have a more definite compass. How narrow should it be? Somewhere, presumably, in between the universe and Missoula, Montana.[1]

A nation provides a fair scope for ethics. Of course, the notion of the nation has had a bad press lately. Nations are oppressive to regions and obsolete in a global era, we have been told. Here, as so often, we tend to be the spoiled beneficiaries of our ancestors. Building this nation took great resourcefulness, a combination of fortitude, ingenuity, and good judgment. It was built, to be sure, on the destruction of Native American culture and the subordination of African Americans and women. But in the end, it made for the inclusion of all and over the expanse of a continent.

The great virtue of a nation is that the people comprising it take responsibility for one another and for what they have in common. When people in Mississippi made it impossible for African Americans to vote, we did not write it off to the amusing or regrettable customs of Mississippians but rather saw to it that they changed their ways. When Alaskans want to drill for oil in the Arctic National Wildlife Refuge, we do not let them go ahead in their state as they see fit but prevent them from disturbing land that is pristine and belongs to the entire nation. This is all to the good. But we have to do more. We have to take our mutual and inclusive responsibility more seriously and understand it more deeply.

"Taking responsibility" may sound meddlesome and intrusive. When a guest of yours begins to take responsibility for the appearance of your home, criticizing the arrangement of your furniture, lecturing you, and moving things around, you will say: "Please don't take responsibility." But if he absentmindedly lights up a cigar and scatters ashes all over your floor, you will say: "For heaven's sake, do take responsibility." Or if he starts to treat a Native American guest with condescension and contempt, you will take responsibility for his behavior and ask him to leave. "Taking responsibility" in a patronizing way is clearly unacceptable. But taking responsibility for what we obliviously and perhaps detrimentally do to one another is recognition or realization rather than intrusion.

Is it possible to say something coherent and substantial about the norms and values that people in this country observe or ought to follow? Isn't this one of the most diverse societies in the world? Consider the cleaning woman in New York City who makes $20,000 a year. The businessman whose office she cleans makes $20 million. She speaks Chinese; he speaks English. She has four children; he has one child. She honors Confucius; he is Episcopalian. Between these two distant points on the social map, there are many grades of income and prosperity, a diversity of religious beliefs and ethnic origins, countless languages—what do all these people have in common?

The cleaning woman and the Wall Street financier share a vision of the good life. She understands very well what he possesses in fame and fortune, and she is determined to help her children attain it. He knows how she lives, and he takes satisfaction in knowing he is better off. There is mobility in society, yet not as much as the rich like to claim to give their status the glow of hard-earned merit and less than the lower classes imagine so they can allow hope to prevail over realism. What social and economic mobility there is, at any rate, builds a highway of common understanding across the thickets of diversity, and hence the janitor could easily do in her leisure what the banker does if only the janitor had his money. She could climb into a private jet as easily as she now climbs onto the bus. She could enjoy the five courses at Le Cirque as well as she eats the hamburger she gets at

McDonald's. It is no longer the case that to belong to the upper class a young woman has to be able to play the harpsichord, do needlework, and remember the steps and bows of the minuet; nor does a young man have to know how to fence, to ride, and to read Cicero in the original.

Obstacles in the Path of Excellence

This country has the right size, and its people share enough of their ways and wants so that we can say something coherent and substantial about their typical conduct and values. But how are we to judge all this? And more daunting still, who could presume to tell Americans how to change their ways if judging their ethics is possible and the judgment turns out to be damning? The answer that leaps to everyone's lips is: No way and no one.

The arguments in support of the answer are second nature to us. First off, in a democracy, it is the individual's right to decide how to conduct his or her life. The only limit on that right is the next person's right to self-determination. This consideration is not only a moral principle, it is a pragmatic necessity, so the argument continues. In a country of so many different convictions and pursuits, telling people how to live their lives would lead to friction and unrest. Finally, if somehow all of us agreed to search for a better moral life, where would we find it? What would it be? Is not the good life a matter of preference? And is it not true that there is no disputing of tastes?

Self-determination or autonomy is definitely a moral landmark of contemporary ethics. It gives our lives spaciousness. We would find life without it oppressive and unbearable. But self-determination is always more narrowly constrained by factors other than equality—the recognition that my self-determination ends where yours begins. Among others, it is constrained by Churchill's principle. In 1943, when the House of Commons had to be rebuilt due to Nazi bombing, Winston Churchill reminded the Members of Parliament: "We shape our buildings, and afterwards our buildings shape us."

The individual does not shape buildings. We do it together, after disagreements, discussions, compromises, and decisions. We, the citizens of this country, through the federal government, were of different minds about interstate roads. We discussed this issue, compromised on the legislation, and in 1956 finally decided that we would build the Interstate Highway System. Once it was built, individuals, so we thought, would decide whether or how to use it. But as we can now see, the possibilities that the system opened up induced people to behave in certain ways. They bought more cars, abandoned public transportation, moved to the suburbs, forgot about sidewalks, blighted the inner cities, drove to Disney World on their vacation, gained weight, and spent a lot of time alone in their cars.

The ways we are shaped by what we have built are neither neutral nor forcible, and since we have always assumed that public and common structures have to be one or the other, the intermediate force of our building has remained invisible to us, and that has allowed us to ignore the crucial point: We are always and already engaged in drawing the outlines of a common way of life, and we have to take responsibility for this fact and ask whether it is a good life, a decent life, or a lamentable life that we have outlined for ourselves.

Inevitably, our common building also forces us to overcome dissent and limit diversity. The way we have done this has led to regrettable losses of cultural variety. The multicultural complexion of American life is more subject to danger than a source of it, and it is not only the constraints from without that imperil ethnic traditions but as much the withering from within that they suffer when faced with the glamours of mainstream prosperity. That internal weakening often leads to terminal subversion where a minority grievance is no longer a plea for the preservation of a tradition or for the right to a distinctive way of life but a brief for a larger share in the standard kind of affluence that is the death of profound cultural diversity.

We have to be fair when it comes to judging the kind of life that has been the result of our shared building and common desiring. Ours is a decent society. But it has troubling features. The common lack of knowledge of physics, biology, geography, history, and politics is embarrassing. Average health is declining and physical fitness is poor. Civic engagement and personal relations are ailing. Knowledge and command of music and the arts, whether popular or elite, are stunted. Awareness of Churchill's principle is dim. Public support of the poor in this country and around the world is the most miserly among the industrialized countries. Our stewardship of the environment is indifferent. The public realm of this country is busy and messy, and most of the public places of recreation and celebration we owe to our great-great-grandparents.

Some of these failings are of a private and personal kind, problems an individual's resolve could deal with, beginning tonight. But a person would do so in the teeth of the larger shape of society. To take the problem of health and physical fitness, there is helpful information, wholesome food, playing fields, running tracks, and enough time for people to eat well and exercise and even become good at tennis or softball. But all these promptings of the good life are swamped by the superabundance of fast and convenient food, by the easy affordability of television and the availability of alluring electronic entertainment right within one's four walls. It was not my decision to build a Hamburger King five minutes from my house or to establish an automobile industry that makes a fine car for half of my year's wages. I did not sponsor research on plasma screens, nor did I organize the writing and staging of witty and captivating television programs. But here I am, surrounded by a cornucopia of tempting food and ready entertainment. Rousing myself to cook dinner, calling my beloved to the table, putting on my coat after dinner to take them on a walk, all this seems forbidding and pointless, given the convenient alternatives.

The troubling features we share in the public realm are simply the lamentable outside of the deplorable inside. In order to put all the consumable treasures of my home within easy reach, the public realm favors utility—transportation links and shopping facilities along with the utilities to support them. The trend to push production, consumption, and affluence, in turn, makes us forget the poor and neglect the environment.

Is there a common root to these issues? Are they symptoms of an underlying malady? Is there a loss of meaning, a decline of values, an end to humanity and history? These are plausible questions, and it makes sense to look for coherence in the welter of troubling features. They are as a matter of fact connected with one another and converge on a central issue. But the center is more tangible and prosaic than the questions suggest, and it is best disclosed by following the everyday leads that point to it.

Not that all is bleakness in this country. American society is not only decent, it can also boast of cultural achievements that are second to none. But they are not our common possessions. If you drive south from Chicago's Loop, you quickly leave the splendors of the Art Institute, the Symphony Center, Soldiers Field, and the University of Chicago and enter an endless expanse of mediocrity and misery. Resignation is the ready reaction, and when challenged and called to account for this sorrowful state of affairs, we are likely to defend ourselves with reminders that the pursuit of excellence is elitist, or oppressive, or a wild goose chase.

The Reality of Moral Standards

The landmarks of decency and the virtues of excellence that are the remedies for the sores on the body politic and the private person are not really in question. The landmarks have been articulated by modern theories of ethics, and they prominently include equality, dignity, and self-determination. While the last of these can be abused to ward off the claims of excellence, the first is at the heart of democracy, and no one can doubt that democracy today is the only viable and vital form of government. The notion of dignity gives equality substance. It tells us what equality minimally requires—everyone's dignity is to be respected and secured. Dignity here is not the distinction that deserves honor and acclaim. That kind of dignity cannot be universal. Dignity as a moral landmark is *inalienable* dignity, the kind that need not be acquired and cannot be lost though it can be violated. It is the reason why we do not tolerate the torture, starvation, or abuse of even the worst criminal. Equality and dignity circumscribe justice in its least requirements—that everyone in this country be fed, clothed, sheltered, educated, and given medical care, and that we make every effort to extend this kind of justice to everyone on the planet.

Regard for the environment is one of the moral successes of the past century. Nothing less than a revolution occurred between 1950 and 2000 when conquest and domination lost their ruling position as the normal approach to nature and environmental concern took their place. We still pollute water and air, we push roads into untouched forests, we drill and mine in pristine areas. But those who initiate such enterprises today can no longer count on popular applause. They now carry the burden of proof that development and exploitation are economically necessary and environmentally tolerable.

When it comes to the individual pursuit of excellence, uniform mediocrity rather than unruly diversity is the problem. Knowing what the norms of personal excellence are is not the problem. There are three reasonable ways that converge on the same three virtues. To recognize the first approach, we have to overcome the vague resentment we feel when challenged to excel. We have been bruised and defeated in our attempts to do better, and yet we know that others have succeeded and that, though often tripped up, we should soldier on. But there is a way of transporting ourselves into a high-minded and generous position. Imagine you are rocking your one-year-old in your arms and the child's fairy godmother appears before you, and she says: "Would it be all right if I were to arrange the course of your child's life so she would become knowledgeable and insightful, well educated in the sciences and letters, and well schooled in judging the character of persons and the circumstances of life? And so she would acquire a courageous heart and become skilled in athletics and brave in facing trouble? And, finally, so she would become devoted to friendship and value her friends and her spouse above all?"

Would anyone reply, No, I want her to be ignorant, timid, and a loner? The three moral skills are the ones that have traditionally been called the virtues of wisdom, courage, and friendship. There is then a second way to excellence, one that began in antiquity and is being followed to this day. Although philosophers today are as reluctant as the next person to tell people what to do (beyond being fair), they do, when musing about the good life, think of it as knowledgeable and insightful and as supported by steady and rewarding personal relations. They are less concerned with physical engagement and prowess, a testament, perhaps, to the nerdy nature of mainstream philosophy.

The third way is that of the social scientist. Much research has been done on happiness, whether it can be validly established and reliably measured, who has it, how it changes, and, important for our purposes, what is conducive to it. As it turns out, being well educated, curious, and informed, being well practiced in meeting adversity and obstacles, and having a warm and reliable marriage or friendship all rank high among the factors of happiness.

Practices of moral excellence flesh out the framework of equality, dignity, and self-determination. They tell us what lies beyond the minimal norms of decency. They begin to give us a portrait of the good person and a picture of the good life, at least in outline, for there are blank and blurry areas. Two are especially notable. Most sketches of excellence show that a life that is blind and deaf to beauty and the arts would fail to be full. So to the three traditional virtues we should add the skill and practice of being engaged in the arts, whether popular or high, as a connoisseur or a practitioner. There is no good name for this virtue, though in the shadows of Aristotle's ethics we find the figure of the "gracious" person, the one who possesses taste, refinement, and beauty and, so I will add, reflects in his or her demeanor the splendor and radiance of the arts and perhaps of religion. Grace, then, would be a name for the virtue of the arts or religion.

The most glaring blank on the moral canvas, however, is the unconcern with Churchill's principle. If we are unaware of how the shaping of our household typically shapes our practices, we can tell our children to do their homework, to stay away from soda pop and snacks, to talk to us, and to practice their instruments till we are blue in the face—it will only create frustration and resentment unless our home is so arranged that doing the right thing comes naturally or at least does not require heroic self-discipline. Here too a tradition and a term for the appropriate virtue are lacking, but again we can wrest them from Aristotle, who was keenly aware of how important the shaping of the household is. He held his nose while talking about this, but he did have a name for the ordering of the household. It is *economy,* and it can serve as the term for being savvy and practiced when it comes to Churchill's principle in the domestic sphere.

Just as personal conduct is shaped and channeled by either the exercise or the neglect of economy, so economy is constrained by the kind of world we have put together collectively. We have a term for the political virtue of caring for equality—it is justice; and there is something of a term for the virtue of caring for the environment—it is stewardship. But what is the term for political rather than private economy? There is in fact a thing called "political economy." Today it refers to the scholarly concern with the connections between politics and the economy. So it is unavailable for our purposes. Again we have to conscript a term that is helpful if not perfect, and *design* has the right connotations. That it refers as least as much to the quality of the things designed as to the virtue of the designers is congenial to Churchill's principle.

Ethics is being equal to the claims of persons and things, particularly to the claims that make us lesser people if we ignore them. The moral landmarks that the modern theories of ethics have discovered and the traditional virtues that set norms for practices of excellence work well in telling us how to treat one another and how to conduct ourselves. But there is this assumption in theoretical and practical ethics that life unfolds on an empty stage, or at least the belief that, when it comes to doing the right thing, the props on the stage of life don't matter much.

That was a reasonable assumption when the material culture changed slowly and its moral significance came to no more than being fair in distributing things and moderate in enjoying things. Justice and temperance are in fact two of Plato's cardinal virtues (the other two being wisdom and courage). But the Industrial Revolution changed the stage of life from the ground up, and now the technological devices that surround us channel the typical ways we behave. Ethics has to become real as well as theoretical and practical. It has to become a making as well as a doing. Real means tangible; real ethics is taking responsibility for the tangible setting of life. Real also means relevant, and real ethics is grounding theoretical and practical ethics in contemporary culture and making them thrive again.

American Ethics

All right then, we need real ethics. But why American ethics? Haven't we had enough grief from the global community for being self-absorbed and overbearing? There are two good reasons why limiting ourselves to the United States is reasonable. First, for an ethics to be relevant to people, it has to address their particular circumstances. Global ethics would have to be thin, or it would be endless if it tried to be concrete and detailed (though the requirement of global *justice* is as urgent as it is general). Second, though the United States is a young country compared with China and Japan, or France and England, it is culturally the oldest sibling of the global family. It has been the first since the late nineteenth century to move through the stages of technological development and to experience their blessings and burdens. This country was the first to have an automotive economy, an inclusive telephone system, a television culture, and cyberlife, and it may again be the first to live in a genetically modified world.

American culture is spreading around the globe. But is American ethics traveling with it? Ethics is sometimes used descriptively. In that sense, it describes the typical conduct of a person or a people. So understood, a certain amount of American ethics goes along with

American culture. But ethics can also be taken normatively, as a statement of norms of moral excellence. Is there something like American ethics in the normative sense? Are there virtues that we and the people in other nations take to be characteristically American?

Yes, there are two such virtues, generosity and resourcefulness. They have distinguished us, at any rate, when we, as individuals and as a nation, have been at our best. As individuals we are extraordinarily generous in providing time and money for good causes. As a nation, we were generous in helping Europe to rebuild after the Second World War and in coming to the aid of Kosovars when they were raped and murdered. There is a more informal and pervasive sense of generosity in this country. You see it in the way we accept immigrants, open our doors to strangers, welcome diversity, and cherish freedom of expression.

Resourcefulness too is a many-sided virtue. Part of it is the persistence our predecessors showed in building a society from scratch (though on the ruins of earlier communities) and in the kind of adversity that is hard to fathom today. Resourcefulness is the ingenuity that was provoked by new circumstances and the convergence of different cultures. It is the refusal to take no for an answer and the readiness to take on daunting projects.

Generosity is the characteristic way in which we have fused the virtues of friendship, grace, justice, and stewardship. Resourcefulness is the American fusion of the virtues of wisdom, courage, economy, and design. To the benevolent observers of this country it has often seemed that what largeness the American soul possesses is due in part to the largeness of this continent. If there is this correspondence, then, to the way technology as a form of culture has shrunk this country, there is a corresponding tendency to let generosity and resourcefulness shrink as well. Of course, the virtues of this country have always been imperiled, and sometimes they have crashed and burned, but the danger that now besets them is unusually subtle and difficult to counter.

CHAPTER TWO

Decency and Passion

Decency

When you look at the moral landscape of this country, you see a broad and basic decency; swirling above it, you hear the rumblings and ragings of moral passions. As for decency, we are, at least individually, generous and law-abiding people, more generous, in fact, than the citizens of most other countries, and not just law abiding, but good-humored and sensible even though our good sense takes time to bestir itself and right the wrongs of indifference, injustice, and imperialism. In everyday life, when the carefully laid plans of two hundred travelers are suddenly disrupted by a cancelled flight, what do people do? Almost to a person, they stand patiently in line to find an alternative flight. When a president resigns in disgrace or is subjected to impeachment, when a presidential election hangs in the balance for weeks, when the stock market tanks or gasoline prices shoot up, you would not know it from the way people behave. They continue to show up for work, give the correct change, stop at red lights, serve on juries, and pay their bills. There is a commendable solidity and regularity to our common life.

In a more formal and stringent sense, we fail to be a decent society. Avishai Margalit, in a finely crafted book, has defined the decent society as "one whose intentions do not humiliate people."[1] The litmus test of such decency is punishment, and this is a test we flunk too often.[2] Our laws of punishment hold much vindictiveness; the conditions of incarceration are dangerous and humiliating; and capital punishment must be the ultimate indignity. But more sensible views are slowly being reasserted.

Movements that are passionately devoted to some cause overshadow the basically placid and solid disposition of society. The degree of passion is usually in inverse proportion to the amount of power a group of people possesses.[3] Between 1989 and 2001, especially, the union of power and passion was hard to come by. Prior to 1989, there were moral issues that allowed anybody to be vociferous and unrelenting. Those issues were the evils of communism, namely oppression, torture, murder, aggression, persecution, and more. That seemed to be the last time the powerful could also be passionate, and they indulged their passions not only for the honorable reason of denouncing evils, but also because they felt morally disquieted by the egalitarianism (more pretended than real) of the communist countries.

Terrorism has given the union of fervor and power another chance. We must defend ourselves, of course, and we must call a criminal a criminal, but the moral ardor of the rich and powerful sounds off-key. Our response, technological and military as it needs to be, should not be so exclusively. There have to be gestures of understanding as well. As it is, the ethical clarion calls, provoked by terrorism, are mostly belligerent and often in the service of maintaining power.

Now that the communist threat has evaporated and as soon as a sense of security has been regained, moral passions and effective power will again become inversely related. Even today, outbursts of moral passion are usually signs of political or economic powerlessness. The powerless are outraged, the powerful are dispassionate. You can see this pattern when citizens complain to the city council, when environmentalists attack the Forest Service, when unions comment on management, when faculty rake the administration over the coals, or when countries in economic distress disagree with the International Monetary Fund. Power and passion vary inversely, and, speaking more precisely, their product in this society is a constant—the more you have of the one, the less you have of the other, but never nothing at all of one or the other.

However, if, as a nation, we want to take responsibility for one another, we have to go beyond impressions and try to determine more precisely what sort of people we are, whether we are morally decent or not. One way of avoiding these questions is to plead the endless complexity and diversity of the population in this country and to counter every assertion about the moral character of the United States with counter examples, exceptions, qualifications, and skeptical reservations.

Social scientists and moral philosophers may find comfort in such skepticism. The former will be rid of the sticky ethical questions that defy empirical answers. The latter don't have to worry about messy and shifting social conditions. But social science without ethics is aimless; ethics without social science is hollow. In fact, the two fields inevitably overlap. There is no social science research that is not tethered, however indirectly, to concerns of social justice and human flourishing, and there are no ethical reflections that fail to appeal somehow to the actual human condition.

Still, there is the theorist's nightmare, not only in social theory but also in the physical sciences—reality might turn out to be so unruly that no widely illuminating theories are possible.[4] In the social realm, to be sure, conventional wisdom sees large and sharply etched features in our society. It is severely polarized and divided by passionately held convictions, so goes the claim.[5] Divided we are, but politicians and the media make it appear as though we were deeply divided with most of the population pushed toward the liberal and conservative extremes. On a graph the line that shows population over the left to right spectrum would have the shape of a U or V. In fact, as Morris Fiorina has put it, we are closely, but not deeply, divided. The political-orientations-over-population curve is bell-shaped; it shows a mountain rather than a valley with most people positioned at or close to the center.[6]

Centrist moderation is a defining feature of basic American decency. Most people are tolerant of others' moral views, favor affirmative action, support the right to abortion with some restrictions, support environmental protection, are skeptical of the power of corporations, oppose job discrimination of gay people, support gun control, favor a guarantee of adequate housing, and endorse universal health care, adequate financial provisions for retirees, and a decent standard of living for all.[7] American decency is also evident in the Transparency International Corruption Perceptions Index. It ranks 146 countries according to how free of corruption they are. The United States ranked seventeenth in 2004, one step up from 2003, not a stellar ranking, but certainly a decent one.[8]

These decent American people seem unsure, however, of their moral force in society. Many think the moral climate has become worse, and yet they can't rouse themselves to concerted and effective action.[9] American decency is distracted and indifferent. It has no vision and no voice in the public sphere. It is largely uninformed and unconcerned. There is no real conception of the good life that decent people are willing to assert and support in political campaigns and elections. Hence the large centrist electorate allows itself to be entertained, aroused, and pushed around by the warring extremists.

Being uncertain of their convictions and unwilling to inform themselves, many voters during presidential election campaigns try to get a fix on a candidate's personality on no more solid a basis than a gut feeling. Trying to get an intuitive sense of a person's character is not necessarily a dishonorable or mistaken way of making a decision as long as the person's character is not an artifact of shrewd propaganda.

Three causes lack power and provoke passion: social justice, the environment, and abortion. The supporters of these causes have never commanded the effective power of government. They find themselves in the roles of petitioners who now and then score victories and at other times try to defend what territory they have been granted. They differ in crucial ways, of course. Social justice deserves our best efforts, the environment some, and foes of abortion none. But in addition to relative powerlessness and passion these movements have something else in common. None of them is animated by a vision of the good life. Their moral visions are narrow, and their language is always in danger of getting shrill, though all of them have moderate advocates with honorable motives. They differ, however, as regards the moral weight of their cause. Social justice as such is a sacred obligation. Environmental stewardship is a significant moral task. Opposition to abortion is for the most part morally confused or morally mistaken though thoughtful concern for emerging human life is an honorable concern. All three causes, however, are likely to be hijacked and represented in the public conversation by extremists.

If the extremists are loud but ineffective and the centrists numerous but silent, what shapes our society and propels it on its way? There is a dynamic pattern to our culture that is as definite as it is invisible. The tacit and implicit support of most people keeps it moving; the clamor of the extremists and the diffidence of the centrist majority keep it from getting on the public agenda.

John Dewey recognized it as early as 1926, and he saw that it was not only a force of cohesion and propulsion, but also that it was the cause of people's bewilderment. In lectures, published a year later under the title *The Public and Its Problems,* he called it "technology" or "the modern age." When "local town-meeting practices and ideas" were swamped by the emerging gigantic national state, technology stepped in to provide organization and coherence: "Our modern state-unity is due to the consequences of technology employed so as to facilitate the rapid and easy circulation of opinions and information, and so to generate constant and intricate interaction far beyond the limits of face-to-face communities."[10]

Technology preempted ruinous individualism, but at the same time "the machine age in developing the Great Society has invaded and partially disintegrated the small communities of former times without generating a Great Community."[11] There is no public, by which Dewey meant a collective moral agent. Technology furnished a supporting and unifying machinery, and it left people apathetic and distracted. "The increase in the number, variety, and cheapness of amusements," Dewey noted, "represents a powerful division from political concern." What Dewey had in mind were "the movie, radio, cheap reading matter and motor car with all they stand for."[12]

The crisis that worried Dewey has materially worsened, but Dewey's analysis has been largely ignored. Philosophers and educators particularly were much more taken with Dewey's proposal for a cure, a pragmatic and experimental approach to social problems. A vision of the good life and the good society, however, does not emerge from experiments alone. To be equal to Dewey's bequest we first need to understand technology more incisively. It is, after all, not a force in its own right that has overwhelmed us, but a certain way of doing things that we have found attractive and in its way effective. And second, we have to locate the places where face-to-face communities, and at length the moral community of this country, may begin to prosper again.

Social Justice

The liberals are my people; universal health care, more equal educational opportunities, subsidized housing, full recognition of women, ethnic minorities, and homosexuals are causes I support; and I would not propose a moratorium on pushing these issues until they have become strands of a richer moral fabric. Still, without such a context, what the liberals have to say to those who have the power and the votes to make liberal ideals become reality consists mostly of scolding and worse. Before 1989 you could hear many a leftist predict that, absent reforms, the masses would rise. The hypersensitive reaction of some of the rich and powerful to anything that faintly smelled of socialism or communism and their relentless suppression of such odors here and abroad are evidence that they were in no mood to ignore those warnings. Now that the imagined specter of insurrection has lifted, the moral question *why* those poorly off should be helped has moved to the foreground, but answers have not. Why help them? Because their ancestors have been wronged? Because they are unwilling or unable to help themselves? Neither of these answers has been morally compelling, and our welfare policies have often rejected them outright.

Uncertainty about the genuine springs of the struggle for social justice has sometimes led to a narrow view of what social justice requires. "Welfare" at first blush looks like an appropriately comprehensive and impartial goal. To provide welfare is to give people what they need to feel well. Critics have replied that this is a blinkered notion of help. A homeless person may feel well when given a bottle of fine whiskey daily. A lazy and shiftless individual may feel well if given food, clothing, and an apartment with a TV set. A teenager may feel well once he has his own car. Critics have urged a broader and more nuanced view of well-being that considers the capabilities of people and furnishes what a particular person requires to develop those capabilities.[13] Yet here too the shallowness of vision harries the pursuit of social justice. Anyone is capable of dozens of things. Which capabilities should we support? The easy answer in theory, though not, alas, in practice, is to stress basic functioning and its requirements of food, clothing, shelter, and health care. Should we include education? If so, how much? When it comes to basics, the welfare approach does no worse than its rival, the capabilities approach. It is an axiom of contemporary ethics that the urgent problems have an ethically straightforward answer and that the subtle and difficult problems lie hidden in the supposed enigma of the good life. Whether to feed the hungry is not a difficult moral question, though getting food to them may be a difficult practical task. How much prosperity the good life requires, to the contrary, is not a question that must be answered immediately, but it is one that arises when people are well fed, and it truly taxes theory.

Amartya Sen, the founder of the capability approach, has cheerfully accepted "the 'incompleteness' of the capability approach—both in generating substantive judgments and in providing a comprehensive theory of valuation."[14] As individual theorists we can exercise this kind of restraint, but as a society and in practice we inevitably favor some capabilities and stunt others. So which capabilities should we favor? To answer the question we need a conversation about the good life. Capabilities are neither here nor there until they are realized as the practices of a good society. Lacking a vision of the good life, we are leaving the fate of the poor to the fruitless disagreement of the liberals who insist on rights and the conservatives who offer market incentives, garnished with compassion. Most of us side with the conservatives and think that only "the deserving poor" are entitled to compassion and help. The liberals argue that the basic welfare of human beings should not be left to the vagaries of the market and fine feelings, but should be guaranteed as a right that comes simply with being a person. Thus the poor are left with a "right" that has few supporters and a compassionate capitalism that is fickle at best. As a result we, as a society, treat our poor more hard-heartedly than most industrial democracies treat theirs.[15]

Environment

The liabilities we find in the moral discourse that supports social justice we find aggravated in environmentalism. I must stress that we and our environment would be much worse off were it not for the work of environmentalists, and yet, what motivates their work is unclear, and had not the environmental movement spawned a cadre of paid professionals and a supporting bureaucracy, that lack of clarity might have gutted environmentalism.[16]

Again, no profound or sublime theory is needed if there is a clear and present danger. Environmentalists sometimes warn us that the survival of the human race is at stake. But 99.999 percent of the human race could disappear and still leave a gene pool and breeding stock sufficient to assure the continued existence of the species. This reply is ludicrous and reveals that environmentalists are not worried about *Homo sapiens* the way they are worried about the bull trout and the mountain caribou. They are moved by a deeper concern, and it is evident in one of the crown jewels of environmental politics—the Endangered Species Act of 1973. It is intended "to provide a means whereby the ecosystems upon which endangered species and threatened species depend may be conserved" and "to provide a program for the conservation of such endangered species and threatened species."[17]

Still, the language of warnings and threats has been the preferred idiom of environmentalists. Many of the dire predictions of the past have been refuted in the event, and this has shown that the culture of technology is more self-righting than self-destructive.[18] The economist Julian Simon was so fed up with environmental doomsaying that in 1980 he offered a bet: One thousand dollars of whatever resources, he claimed, would cost less in ten years—demonstrating that the world was not running out of resources. Whoever took the bet would, ten years hence, have to pay the difference between the lower price and $1,000, or Simon, if he turned out to be wrong, would pay the difference between $1,000 and the increased price of the resources. Paul Ehrlich, a prominent doomsayer, took the bet and proposed to invest $1,000 in five metals. By 1990, all of them had become cheaper, and Ehrlich had to send Simon a check for $576.07.[19] But while things here on earth have improved ecologically, global warming has overtaken us from above. Warnings are again in order. Still, what if the community of nations agreed on a policy that would slow global warming and eventually halt it, would this be the end of environmentalism?

Mainstream environmental philosophers have tried to find a foundation of ecological concern that would remain morally unshakable whether the planetary environment will be secured or remain endangered. That bedrock is "the intrinsic value of nature." As Baird Callicott has it, "providing theoretical grounds for according intrinsic value to nature (or some of its parts) is a worthwhile project—the principal, the defining project of environmental ethics."[20] There is variety and subtlety in the various arguments for the intrinsic value of nature.[21] What they have in common, nevertheless, is the attempt to cordon off nature from the trespasses of humans. To do this physically is often the only way of saving a species or ecosystem. But doing it conceptually and theoretically is a dubious enterprise.

The intention is plausible enough. If philosophers could establish the intrinsic value of nature to the satisfaction of a majority of people and get the majority's compliance, wild and free nature would be safe. But even if successful, this project goes too far and not far enough. It would accomplish too much in establishing nature as valuable regardless of human beings. This would make nature not only immune but also indifferent to the works and ways of human beings. Such immunity is by now an illusion, and such indifference would come to an impoverishment of the human condition. As Bill McKibben has shown so mournfully, nothing on or above earth is any longer beyond human intervention.[22] Philosophers typically think of the wilderness as having intrinsic value, and by "wild" they mean pristine and untouched.[23] But nothing is that way anymore. Wilderness areas today are shaped by the decisions humans make about encroaching exotic weeds; about fire suppression and hunting regulations,; about the reintroduction of wolves, grizzlies, fishers, and lynx; about acid rain; and of course about global warming.

One reaction against the arguments for the intrinsic value of wilderness has been the radical embrace of human agency and the contention that the very notion of wilderness, whether it refers to untouched or to tainted nature, is a human construction. A celebrated example of such environmental constructivism is William Cronon's "The Trouble with Wilderness; or, Getting Back to the Wrong Nature."[24] Cronon's language abounds with phrases that identify the wilderness as "a human creation," "a product of civilization," "entirely a cultural invention," or "this complex cultural construction." We have been "freighting it with moral issues," he says, and it has "become loaded with some of the deepest core values of the culture."[25] Such radicalism has a fine time until it comes to making decisions. The intrinsic value school has at least one clear conclusion to draw from its position: Stay out! But if it is all construction, only human whimsy and power remain as courts of last appeal or would remain if constructivists did not in the end fall back on realism.[26]

That establishing the intrinsic value of nature is not enough appears from the abrupt and negative instruction it is confined to—Do Not Enter, or something similarly forbidding. What needs to be shown, in keeping with the most profound environmental concern, is how nature regains a new moral voice in the midst of a technological era and presents itself and speaks to us not only in the distance and from afar, but in our civilized and urban conditions and tells us that we must change our lives.[27] Philosophers do not have jurisdiction over the status of nature. They have to be nature's trustees and advocates.

Abortion

Environmentalism is all sweetness and light when compared with the moral passions of abortion opponents. Their fury has notoriously employed the language of threats and gone beyond that to harassment and murder. Most right-to-life proponents deplore violence, but some of them foment the atmosphere of lethal anger by calling abortion providers murderers and likening their work to the holocaust. Legislators have made abortion burdensome through the well-acronymed TRAP laws—targeted regulation of abortion providers.[28]

The ostensible motivation of abortion opponents is commendable enough—the protection of the lives of human beings. The great difficulty is to determine when in the course of pregnancy there is something like a human being.[29] It seems intuitively wrong to claim that there is one from the moment of conception. A clump of cells is just too distant from what we know a person to be. It is much more natural to think that in the growth of the embryo and fetus a human being is developing gradually or in stages. This, at any rate, is the view of Thomas of Aquino whom Catholics in all other matters of doctrine regard as the Angelic Doctor.[30]

The confusion about the springs and sources of the right-to-life movement is particularly thick in the case of the many partisans who support abortion in cases of rape or incest. In the eyes of true believers such as the Catholics this amounts to killing an innocent "baby" simply because of the terrible circumstances of conception that the "baby" cannot possibly be responsible for. If someone needs to be killed, should it not be the rapist or the incestuous relative?

But perhaps benign inconsistency is preferable to the narrow and erroneous rigor that is the official doctrine of the right-to-life people though at times the inconsistency is breathtaking. Finding "no necessary logical inconsistency here," political scientist Morris Fiorina reports: "But—and this is a critically important—not everyone who believes that abortion is wrong—not even everyone who believes it is murderous—supports making it illegal."[31] To be fair, narrowness and error can also be found in the ethics of some of the free-choice advocates. They seem to believe that an individual's choice trumps all claims that someone else may have on that individual and that sometimes pregnancy is simply a nuisance to be rid of.

In fact, however, conception is a momentous event that has brought a living being into existence, a creature that, absent grave reasons, has a right to life. But what is a grave reason? This question is as difficult as the one about the emergence of a human being. In the end only the pregnant woman can answer the question. No doubt some pregnancies are incurred and ended superciliously, and legislators are right in taking exception to them. But they are wrong in thinking a law can pick them out and even more wrong in assuming the law could provide a remedy for such irresponsibility. The law is a very blunt instrument and causes grief and sometimes the mother's death when loosed on the complex circumstances of pregnancy. Some circumstances are more than complex; they are miserable and desperate, and ready access to abortion is the only way out. It is misery and despair that lawgivers need to keep in mind without requiring proof of misery and despair. To demand such evidence is in many cases to exacerbate the calamity.

It must be a welter of good intentions, resentments, and confusions that fuels the abortion movement. The consequence of all this has been the contraction of complexity into a narrow all-or-nothing conception. If an egg and a sperm have united, there is a human being; if not, not. This simplistic notion of human development is used as a substitute for a vision of virtue and moral excellence. The simplicity of the issue, the obvious identities of the opponents, and the sanction of revered spiritual leaders give the zealot an energizing and feverish purpose in life.

A moral value has this in common with a gas: when it's compressed, it gets hot. But when so compressed, a moral value, no matter how heated, takes little space in one's life. It fails to inform what we do daily, and it makes little difference to the life of a true believer whether that narrow value prevails or not.[32] Consider the foes of abortion. At least half of them are men, and whether abortion is legal or not does not touch them immediately. Neither does it touch the female opponents who are not of childbearing age. Of those who are, a majority is in a position to control their pregnancies. And so on down to the relatively few for whom abortion is at a particular moment an utter and bitter necessity. Similar considerations hold for proponents of the death penalty and opponents of gay marriage.

The opposite can happen also. Moral passion gets diffused and pointless. The supporters of social justice and a healthy environment have sometimes been gathered and moved by a free-floating anger that issues in protests against corporations, capitalism, world trade, and globalization. There are, to be sure, fragments of genuine concern for the exploitation of workers and about damage to the environment. But these shards are mixed up with resentments and fail to yield a coherent view of the future. Here is fury without much doctrine and little focus.

Moral passion, whether compressed or diffused, reveals a profound need for a moral vision, a need that deserves respect. But manifestations of such need require examination as well. For now, let conjecture stand in for explanation. The ground of contemporary culture must be so compacted and barren that a rich and grounded moral vision has a hard time taking root and gaining public support. It's the moral complexion of contemporary culture, then, that needs to be investigated and reformed.

Without a more grounded and fruitful ethics, moral passion in both its narrow and wide versions can turn violent and even criminal. Abortion providers are attacked and murdered. Global conferences and corporations are threatened and vandalized. But degree of passion is in the end inversely related to effective power. The dispassionate establishment has law, order, and the police on its side.

This is an arrangement we accept, and we are wise in doing so. Those who disagree with it and in protesting it violate the law are in most cases willing to appear in court and take their punishment rather than retreat to a Montana ranch and defy the law come hell or high water. In general, we submit to changing the law in lawful ways.

CHAPTER THREE

Kinds of Ethics

Legality, Morality, Civility

We have worked out a system that appears to exhibit a fine division of moral labor between laws and ethics. The rule of law is *universal, compulsory, and morally minimal*. Everyone is subject to the laws, the laws are enforced, and the laws refrain as much as possible from telling us how to live our lives. Of course, compliance with the rule of law is not enough to lead an ethically commendable life. An entirely law-abiding citizen may yet be crude, selfish, and lazy. But neither is the law the only guide of conduct for most of us. Typically we have embraced moral rules that in part overlap with the laws but in large part also exceed them.

For many people in this country, religion provides the rules of conduct that go beyond the requirements of the law. Private morality is in any case *particular, optional, and often maximal*. While the laws hold for all citizens, private morality pertains to a particular group of people. While compliance with the law is compulsory, obedience to a particular morality is optional—at any time you can opt out of a moral community, and the government guarantees your right to do so. And whereas the moral weight of the laws is to rest as lightly on us as possible, moral communities can make large and sometimes supreme ethical demands, namely to forgive one's enemies or to share one's possessions with the poor.

Hence the apparent division of ethical labor between legality and morality. The laws take care of the dealings all of us inevitably have with one another. The various kinds of morality govern, more or less, our private and optional enterprises. It is a division that characterizes and is sanctioned by liberal democratic theory. It is reflected in the separation of church and state and in the homely wisdom that tells us: You can't legislate morality. The legal part of this arrangement is thought of as the sturdy foundation of life, the moral part as a cloud of balloons that hover inconsequentially if colorfully above the solid fundament that alone matters. A competing model sees the several moralities fighting it out like pit bulls in an arena made secure by the law.

The official view that sees a sharp division between laws and private moralities has to be wrong, for it fails to account for the decency of our common moral life. Private morality is strong in its ethical instruction, but it divides into particular and limited communities. The laws are common to all of us but morally too weak to foster common comfort and tranquility. They are simply insufficient to guarantee a flourishing society. There is no law to keep one from being uncooperative, unreliable, or rude. The law puts limits on some of these vices. Thus you cannot refuse to pay taxes, disregard a contract, or intentionally trip up people. Within those limits, however, one can lead a life that is nasty, brutish, and legal.

Since our society is normally decent and often commendable, there must be a force that is morally more powerful than the laws and more inclusive than the various religions and philosophies of life. We can call it civility.[1] It lies halfway between legality and morality. It is *common,* that is, less than universal; it is *encouraged* rather than enforced; and its requirements are *morally demanding* though not supremely so.

Civility is needed not only to make our common life agreeable and enjoyable, it is needed as well for the effectiveness of laws. Violation of the laws is punishable, of course, and the police power of the state forcibly descends on scofflaws. But fear of punishment is insufficient to insure compliance with the laws. If citizens individually and in groups regularly calculated the gain from breaking a law against the risk of being caught and proceeded accordingly, life would quickly become unlovely. The goodwill and trust of civility are the lifeblood of law-abidingness.

Just as civility infuses legality in a decent society, so morality informs civility. Philosophies of life, religions, associations, and clubs all have their special and exclusive concerns. But they also have, often in consequence of their particular convictions, teachings of charity, forbearance, and discipline that they share with other moral systems and that contribute to civility. Thus rather than thinking of our moral structure as consisting of the crucial legal basis and the optional moral adornments, we should see it as a watershed where the springs and sources of morality feed the loosely bounded stream of civility that in turn flows into the clearly marked channel of laws and regulations.[2]

As the exhaustive and impressive work of Robert Putnam has shown, the springs and principles of civility seem to be growing weak. One can draw two conclusions from this: Putnam's who thinks that this drought will imperil our health, happiness, security, justice, and prosperity, or the still more melancholy conclusion, given our undiminished tranquility and growing prosperity, that the fresher and deeper currents of civility have become dispensable, replaced by an ever more sophisticated and stable technological machinery.[3]

To consider every abortion a misfortune is perfectly legitimate, and to promote caution and self-discipline through persuasion and example is not only legitimate but also morally commendable. However, such endeavors fall into the middle ground between legality and private morality, that is, civility. Conservatives are not alone in overlooking or mistrusting civility. Liberals have tried to outlaw politically incorrect speech on campuses. Such speech is often morally repugnant. But when it occurs, we should not call for the campus police or administrative intervention, but instead denounce such bigotry publicly and rally in support of those who have been defamed.

Theoretical Ethics, Practical Ethics, Real Ethics

We find confusion and narrowness not only in the pleading of moral causes but also in the perception of how the regions of moral conduct are bounded and how they hang together. The remedy for these ailments is ethical reflection. More particularly, ethical theory is the medication of choice for confused or indistinct problems and principles. Consideration of ethics as a practice will broaden the focus to take in the richness and texture of the good life. Material reality finally, needs to be examined because it channels the round of daily activities in ways that make or break the nobility of our lives. It is in the quality of the material culture where the destiny of the good life and the good society comes to completion.

Just as the regions of moral conduct get confused, so do the kinds of moral reflection that are supposed to cure confusion. How do *they* hang together? The first thing to get clear about is an ambiguity in what is meant by theory in ethics. In a crucial sense, all ethics done by Western philosophers is theoretical. It always consists of reflection, argument, criticism, refutation, and persuasion. It does not contain, for example, practical exercises in sitting, breathing, and meditating as do certain Eastern philosophies that are admirable in their own right. Theory in the broad sense of reasoned reflection and intellectual argument is the typical tool of Western thinkers. It is not always, however, especially not in ethics, the object of their analyses and proposals. In fact, the distinction between the three types of ethics is based on the different *objects* of moral reflection and argument, not on the tools used in working on those objects.

Accordingly, I call *theoretical ethics* the school of moral philosophy that not only uses theory as a tool but also claims that an ethical theory is at the heart of what it reflects on—truly moral action. Anyone, so the claim goes, who acts ethically, obeys or exemplifies an ethical theory or principle, at least implicitly, and moral conduct gains, so it is argued further, when the theory is made explicit and is followed conscientiously.[4]

Theoretical ethics, as we shall see, has its indispensable significance and its unsurpassable grandeur. But in its claims to exclusive authority, it overreaches. There are two standard versions of theoretical ethics, Immanuel Kant's (1724–1804) ethics of duty and John Stuart Mill's (1806–1873) ethics of happiness. They have lately found a competitor in evolutionary psychology. These three schools of thought direct our attention to the moral quandaries, large and small, that punctuate our lives, and all three instruct us to bring a principle to bear on the quandary and so to effect a solution of the quandary.

This is not to deny the informal moral wisdom that speaks from the writings of Kant and Mill and of Daniel Dennett for that matter. Still, it is a moral principle, applied to definite difficulties, that is placed at the center of theoretical ethics. The question is: Can applications of a principle do justice to the breadth and depth of the human condition? Will a sequence of properly solved quandaries amount to a good life? It will certainly hold a person blameless. The good life, however, extends further and requires more. It is a matter of daily practice, of acquiring moral skills and habits, of keeping them sharp, and of exercising them regularly. The field of athletics, where one trains and keeps in good shape, provides a far better model of the good life than the court of law where incidents are brought under laws.

Martin Heidegger in 1927 and Michael Oakeshott in the 1950s criticized the pretensions of theory to incisive understanding and beneficial guidance.[5] Heidegger did so as regards our daily engagement with the tangible world; Oakeshott made his point with regard to politics. In our time, Charles Taylor and Hubert Dreyfus have argued for the sovereignty of engagement and practice vis-à-vis theory.[6] In the late 1950s, G. E. M. (Elizabeth) Anscombe and Philippa Foot both criticized contemporary moral philosophy along similar lines and revived virtue ethics as a more appropriate way of illuminating and directing moral life.[7] The great ancestor of virtue ethics is of course Aristotle. Twenty-four hundred years ago he warned his students about the inappropriate hankering after precision in ethics, and he developed a moral philosophy that concentrates on moral skills and practice, on the virtuous person, and on the good life.[8]

By practical ethics, then, I mean the confluence of the philosophy of practice and of virtue ethics. There is one more stream that comes under the heading of practical ethics. It is the application of ethical theory to obvious problems of daily life, especially to those arising in the practice of medicine and in our dealings with the environment.

Theoretical ethics and practical ethics are well-established genres of moral reflection. Real ethics is not. The reality it talks about is the visible, tangible stuff that engages and surrounds us. Reality ranges from the homely to the monumental, and it is ethically charged at every level. It matters morally whether you have television in your home or not; and if you do, it matters what kind of television it is, whether a small set with a cathode ray tube and analog picture or a gigantic plasma screen with high digital definition. It is ethically consequential where you place the set, whether in the middle of the living room, in a home theater, or out of the way on the third floor. It is significant where your home is located, whether in the country, in a suburb, or downtown. It matters morally what all is physically reachable and included in your life, whether churches, synagogues, or mosques, parks, tennis courts, or stadiums, museums, theaters, or concert halls, and mountains, prairies, or seashores.

If you look at the titles and abstracts of the programs that the American Philosophical Association publishes for the meetings of its three geographic divisions, you will by and large find supreme indifference to these matters. Philosophers of ethics in particular assume that moral conduct governs, but is not governed by, the tangible environment. There are important exceptions. Philosophers of technology are very much concerned with the shape and import of the tangible culture. But they are marginal within the profession though they have contributed at least as much to the public conversation as their mainstream colleagues.

Theoretical ethics, practical ethics, and real ethics should be thought of not as rivals but as complements of one another. Together they make up something like a complete ethics just as equal-sidedness, four-sidedness, and equality of internal angles make up a square. With any one or two of these complements you can't be sure you'll get a square. Or to use another analogy, theoretical ethics provides the skeleton, practical ethics puts tissue on the bones and furnishes an organism, and real ethics sets the organism in its environment and allows us to see whether the organism is prospering or not.

To summarize, I propose that theoretical ethics gives us the landmarks or the framework of contemporary ethics but, left to itself, gives us an impoverished view of the moral life. Practical ethics attends to the texture and the richness of ethical conduct but, if it goes no further, remains inconsequential and inconclusive as regards the quality of our lives. Real ethics investigates the moral structure of the material culture and thus reveals the levies, dams, and channels that constrain the course of life, and it discloses the things of art and nature that inspire and engage us.

To look at the proposal from yet another angle, we can think of theoretical ethics as raising the question "What must we do?" and in reply directing us to a blameless life and a decent society. Practical ethics raises the more ambitious question "What should we do?" and by way of an answer directs us to a life of virtue and a commendable society. Real ethics, finally, takes Churchill's principle to heart and asks "How should we live?" The response is the endeavor so to shape our world that it inspires the good life and the good society.

PART ONE

Theoretical Ethics

CHAPTER FOUR

Moral Landmarks

Equality, Dignity, Self-Determination

Equality, dignity, and self-determination are crucial to the way we Americans think of ourselves as a moral community. These standards arose at the dawn of the modern period, and they had their first trials and successes in this country. Equality is the first concern of the Declaration of Independence, equality among peoples and males. The exclusion of women and slaves from the principle of equality is a sorrowful fact. But even the limited equality of the Constitution was contested in the early republic and threatened by monarchy. The danger of monarchy was not that one man would grab absolute power and declare himself king; the specter was rather the accumulation of power by a landed and moneyed elite. Thomas Jefferson was the most vigorous defender of democracy, but though he was deeply troubled by slavery and wished to see it abolished he did not promote abolition, nor did he believe in the equality of African Americans with white people. It took a civil war to extend equality to all African Americans and strenuous campaigning in the late nineteenth and early twentieth century to realize it for women. Equality turned out to have a moral force that carried it beyond its initial limits. Today its power is being tested once more in the struggle of homosexuals.

Dignity gives equality substance. The term is not mentioned in the Declaration, though it is implied in the prohibition of cruel and unusual punishment, and the Supreme Court at one time found it to be implied at an equally prominent place—the Preamble to the Constitution. "From its founding," the Court said, "the Nation's basic commitment has been to foster the dignity and well-being of all persons within its border." That commitment, the Court held, was evident in the Preamble's resolve to "promote the general welfare, and to secure the blessings of liberty to ourselves and our posterity."[1]

At any rate, the distinctive quality of dignity is mentioned in the Declaration. The modern norm of dignity is not reserved for dignitaries or dignified personages. Everyone has it by nature. Moral dignity is "unalienable." Dignity defines the terms of equality. We are, after all, quite unequal in strengths, smarts, and looks. But we are all equal in dignity. There is a high-mindedness to how we understand equality. We are not equals as rascals or losers. Rather, each of us has inviolable worth.

Dignity has acquired a characteristically American cast in this country and, again, one that had a hard time becoming established. A sense of prowess and pride was common among the Americans who first settled on this continent. The first white settlers came with a sense of mission and a high purpose. It has taken some four hundred years for the white people to recognize fully and belatedly the dignity of the first Americans. And ethnic immigrants, even after their equality had been established in the abstract, had to struggle, wave after wave, for the recognition of their dignity.

The crucial problem with dignity today revolves around the question: What does a person need to lead a minimally dignified life? Or: What do we as a society have to do practically to meet the theoretical standard of dignity? In theory, the answer is not difficult: Healthy food, decent housing, basic health care, elementary education, minimal income. President Franklin Delano Roosevelt, in the face of fascism and communism, called for a Second Bill of Rights to secure such basic dignity for all.[2] But the myth of rugged individualism first and the anesthetic effect of technological comfort later made us the least compassionate among peer societies.

If dignity lends or ought to lend substance to equality, what are the grounds of dignity? What entitles us to take a high-minded view of ourselves? It is each person's ability to determine the course of his or her life. Each of us is entitled to self-determination. The new world was seen as the land of freedom and challenge where people could escape the bonds of poverty and oppression and make something of themselves. With all the injustice and misery that the European settlers inflicted on Native America, there was a heroic aspect to the destiny that men and women had chosen in coming to this country. Leaving Europe was the liberating part of self-determination; arriving on these shores was the challenging and sometimes crushing part of making your own place and life.

Today, we are chiefly heirs and beneficiaries of the struggles for equality, dignity, and self-determination. In moments of moral confusion, we look to these three moral landmarks to get our bearings. Although they worked their way through layers of oppression and prejudice and now loom large, I am not sure they would be visible to the untutored eye. It took the work of ethical theory to instruct us and to reveal them clearly and distinctly.

Immanuel Kant

Immanuel Kant (1724–1804) provided the theoretical groundwork for modern ethics. It was his ambition to lay down the moral law, a rule of conduct that would be universally and necessarily valid. The clear and explicit statement of the moral law would also, Kant was convinced, put scholarly ethics on a solid footing, and thus put an end to "the disgusting mish-mash of cobbled-together observations and semi-intelligent principles" that, as Kant saw it, dominated the popular writings on ethics of his time.[3] This is not so different from what best-seller lists show today, and what Kant can give us even now is a sharply etched picture of the moral lineaments we can trace in the history of this country.

The classic work of Kant's moral philosophy is his *Foundations of the Metaphysics of Morals* of 1785. It revealed the moral skeleton that has given modern ethics its cardinal shape. Here we find the three moral norms of equality, dignity, and liberty. Kant articulated them as commands. The norm of equality he spelled out as the celebrated categorical, that is, unconditional, imperative:

> Act only according to that maxim whereby you can want at the same time that it become a universal law.[4]

The norm of dignity he phrased as a command that is often invoked in medical ethics:

> Act so that you treat humanity, in your own person as well as in the person of everyone else, always as an end and never merely as a means.[5]

Finally, Kant put the norm of liberty as the injunction of autonomy:

> [Act only so] that the will, through its maxim, be able to regard itself simultaneously as legislating universally.[6]

Kant thought of these commands as three versions of one and the same moral principle.[7] He also thought it impossible to give an explanation of the law. Hence what remained to be said were reminders, clarifications, and defenses of the ethical principle. Today we find the points raised in defense of the third version of the moral law the most congenial in Kant's rich and remarkable treatise. To put the basic insight of that version more simply, ethics requires that we only obey that moral law that we impose on ourselves. Such self-legislation is what the word autonomy means. Freedom for Kant is autonomy.[8] Its significance comes into relief against the other possible reasons why one might submit to moral commands.[9] Of these, two have always been most powerful—desire and fear or, more pointedly, greed and cowardice. But they are likely to provoke rueful reactions—cowardice most immediately because, even while acting cowardly, we know we are debasing ourselves. Kant's rejection of desire or, more strongly, greed causes us difficulty since we have so often indulged our belief, if only in daydreams, that real freedom would be the license and the ability to let our desires run wild. But occasions where fortune or self-indulgence has pleased our desires are often followed by disappointment or even sorrow.

Although Kant attempted to justify the first and second imperatives in their own right, the second really draws its strength from the third and the first from the second. We all sense, and the observers of the modern predicament are explicit on the point, that each of us carries a burden of self-determination that sets us apart from our ancestors and endows us with a special dignity.[10] While in earlier times the crucial moral question often came to "How do I find my place in the established moral order?" the question that haunts us today is

"Which moral order can I responsibly accept?" We realize that any moral regime that has not been chosen *by us,* but claims to answer this question *for us,* is unacceptable.[11]

The burden of freedom as autonomy lends us dignity. We are subjects, submitting to the moral law, only because we are at the same time sovereigns, giving ourselves the moral law. Such is the enduring support for the principle of human dignity. To use a person is to disregard the person's moral sovereignty and so to violate the second principle. Dignity in Kant's sense justifies equality. On the face of it, humanity divides into very unequal persons. Some of us are handsome, others are homely; some are healthy and strong, others are sickly and frail; some are gifted musically, others have a tin ear; some are of African ancestry, others of European. But we are all alike in being moral agents, beings that can and must determine their own moral conduct.

The chain of moral justification does not end here but rather closes on itself. The first principle informs the third. There is a crucial constraint on the moral law we are free to impose on ourselves—"we can choose any rules to govern our behavior, just so long as these rules can be applied to all cases and to everyone," as Gordon Brittan has put it.[12]

The most astounding and influential presentation of equality, dignity, and liberty was the United Nations' Universal Declaration of Human Rights of 1948.[13] These three norms figure prominently in the first sentence of the Preamble and in the first article of the Declaration.[14] They are not the only ones—economic rights, justice, and peace loom large as well. Nor is Kant, historically speaking, the first architect of human rights. But he gave the three key terms a penetrating theoretical treatment. Ethical theory in fact came into its own when the UN's Commission on Human Rights began its deliberations. Kant would have found the claims and arguments familiar. He would have disagreed with those who opposed human rights, and he would have rejected some of the arguments that were made in support. At any rate, he would have embraced the clear-minded, secular, and cosmopolitan vision of the Declaration. The passage of the Declaration by the General Assembly of the United Nations on December 10, 1948, was a momentous event. We can take pleasure in knowing that the United States contributed substantial expertise to this achievement and the widely admired leadership of Eleanor Roosevelt.[15]

The benefit of this accomplishment was that from then on there was a powerful and universally acknowledged language to capture and condemn social and political oppression and cruelty. Without the Declaration, a critic of apartheid or exploitation would always have faced a double and difficult task, first to make a case for human rights and second to criticize injustice on the grounds of the case made. With the Declaration, such critics have ever since had an articulate and authoritative basis on which to proceed. "Ideas," as Paul Lauren has put it, "know no boundaries and have the capacity to change the world."[16]

Diversity and Validity

"Cultural imperialism" was a charge that was leveled already at the supporters of human rights on the UN's Commission.[17] Some sixty years later, we still face the question of whether we should overturn national sovereignty in the name of human rights and democracy—the form of government that alone can secure those rights. Is there perhaps a form of Asian human dignity that does not imply equality and self-determination? If a plurality in a country like Algeria freely votes, as it did in 1991, for a fundamentalist Islamic government, have the defenders of democracy and secularism, in this case the Algerian military, a right to prevent the result?

The last question reveals a conundrum at the very heart of democratic theory: Can a society freely give up its freedom? This used to be a precious puzzle of political theory. It has become a troubling reality and has an analogue in the riddle we find amid the Kantian landmarks. If you have a right to self-determination, then I have a duty of tolerance. I cannot tell you how to live your life. The only limit on your self-determination is my freedom to determine my life. The practical version of tolerance is usually thought to be this: You can organize your life any way you like as long as you do not harm anyone else. It's a good principle, yet it is bordered by difficulties that range from the intriguing to the troubling. Are we harmed by people who decide to walk about in the nude? What if they smoke marijuana? There are gross cases. Can two adults consent to cannibalism?[18] The answer to the first question is yes and to the third no; the second one is more difficult to answer. The arguments in support of answers come from our understanding that common decency is more substantial and important than harmlessness. It is the behavioral counterpart to Churchill's principle. We shape our principles of decency, and afterward our principles shape us. Just as we do not want to live in a rundown city, we do not want to live in a bizarre society.

This way of looking at the substance and limits of tolerance leads to a collision of tolerance and equality. Worse, it sanctions misery and misunderstands ethics. If common morality and its limits of tolerance are things we, as a society, have constructed in the same way we have built our cities, then what are we to do when we deal with a society whose moral construction we disagree with? What about a society where women cannot get an education or own property? Should we be tolerant of the social construction or urge equal rights for women?

If women in that society claim equal rights, the situation is morally clear if politically difficult—we should help them in the best possible way to attain those rights. But what if the women don't? I was once challenged to appreciate the "fact" that among the indigenous people in Chiapas, Mexico, incest was customary and accepted and that women expected a beating from their husbands as an expression of love.[19]

We seem to be caught between the relativism of social constructivism and the absolutism of "The Universal Declaration of Human Rights." Given a choice, postmodern theorists are often drawn to the more provocative alternative, the one that makes their opponents seem timid and conservative. Demanding respect for distasteful features of another culture raises more eyebrows than once more defending equality and self-determination does. A provocative posture in this particular case can also count on rueful recollections of Euro-American chauvinism that used to look down on indigenous cultures, deriding their beliefs, condescending to their arts, belittling their technology, and ignoring their wisdom. We must also recognize that it would be unfair to condemn the women who submit to domestic violence and to convict them of stupidity or cowardice. Evidently equality and self-determination are not yet moral options for every woman in Chiapas. But whether normal or not, wife beating and incest are not uncontested in Chiapas. There are indigenous voices demanding equality, dignity, and liberty for women.[20]

What then about the conflict between Enlightenment absolutism and politically hypercorrect relativism? The answer is that a threefold relation needs to be worked out when we encounter a premodern or foreign culture. To begin, there are always moral norms that we share with another culture, virtues like courage or generosity or social arrangements such as private property or monogamy. Then, unhappily, there are sometimes customs we know to be wrong—human sacrifices, slavery, wife beating, and raping daughters. Finally there are standards and social arrangements that we can entertain as moral possibilities without being able to embrace them, matriarchy, for example, or government by a council of elders (the original meaning of *senate*). Matters are more complex, of course. Ethical ideals and practices are interwoven with one another and cannot be divided simply and neatly into the three categories above. Something like a combination of endorsement, rejection, and empathy will have to shape our relations to moral strangers.

It is often thought that ethical validity and ethical universality stand and fall together, that a moral norm can be binding only if it is binding everywhere and at all times. But there is no such logical tie. Consider democracy. It is the only viable form of government today, but it was not such for Shakespeare. The plays that deal with the rise and fall of kings, the so-called histories, do not challenge the notion of kingship. Shakespeare's question is not whether kingship is an acceptable form of government, but rather what a weak king is, a good king, a murderous king, an ambitious pretender to kingship, and so forth. Some of us may be able to empathize with the validity of monarchy, but no one of consequence would think of embracing any form of government but democracy.

The underlying norms of equality, dignity, and liberty were uncovered at a particular time, and yet there is no question but that they are binding on us. They are gradually spreading around the globe and sometimes clash with traditional forms of privilege, subservience, and bondage. But they must and will prevail. They are binding on everyone today. Thus beyond absolutism and relativism lies what we may call epochalism—the realization that every epoch in history has its uniquely characteristic and valid norms.[21]

The norms of equality, dignity, and liberty are clearest and most authoritative when they are violated. It's this moral fact that reveals the rise and force of epochal moral norms. When we see or learn of women raped, homosexuals derided, children abused, or minorities exploited, it is then, in actual life, that proof of ethical norms is forthcoming. Appeals to male prerogatives, biblical passages, parental authority, or superior breeding sound hollow and disingenuous. When prejudices of old have lost their power, we are witnessing a new epoch. Not that revulsion and repugnance are of themselves infallible and sufficient. Immediate experience and calm reflection need to converse and find equilibrium.[22]

CHAPTER FIVE

Jefferson and Kant

Reason, Ethics, and Ordinary People

Immanuel Kant seems foreign to the history and moral climate of this country.[1] Compare him to Thomas Jefferson whose life overlapped with Kant's by sixty-one years. Kant was a philosophy professor who remained unmarried and lived his entire life in his native city of Königsberg. Jefferson was a statesman, architect, musician, horseman, naturalist, historian, and plantation owner. He was a loving husband, though widowed early, and a devoted father if not such to all his children. He traveled widely in the United States and in Europe. Philosophers like to hold forth on the influence of Kant, but what actual impact he had on German and Western culture is one of the great sociological unknowns. There is no doubt, however, that Jefferson had a strong hand in shaping the beginning of the United States, its geographical extent, its educational system, its architecture, and, to sum it up, its culture.

But both were men of the Enlightenment and were profoundly attuned to the rational and egalitarian spirit of their time. Reason was for both of them the source of light. It figured prominently in Kant's *Critique of Pure Reason* (1781) and *The Critique of Practical Reason* (1787). "'Dare to use your own reason!'—that is the motto of the Enlightenment," Kant said in his essay "What Is Enlightenment?" (1784). He continued: "For this enlightenment, however, nothing is required but freedom, and indeed the most harmless among all the things to which this term can properly be applied. It is the freedom to make public use of one's reason at every point."[2] Similarly Jefferson said in 1790: "It rests now with ourselves alone to enjoy in peace and concord the blessings of self-government, so long denied to mankind: to shew by example the sufficiency of human reason for the care of human affairs and that the will of the majority, the Natural law of every society, is the only sure guardian of the rights of man."[3] With all the common sentiments, there is this difference: Kant addressed a chagrined plea to the elite that read the progressive *Berlinische Monatsschrift*. Jefferson spoke with pleasure and confidence to his fellow citizens of Albemarle County.

Both men, however, put their trust in the moral fiber of common folk. Kant in 1785 said "that neither science nor philosophy is needed to know what one has to do in order to be honest and good, and indeed wise and virtuous. It could have been guessed to begin with that the knowledge of what everyone is obliged to do and thus also to know would be within the reach everyone's, even the most ordinary person's, business."[4] Two years later, Jefferson had this to say about moral philosophy: "I think it is lost time to attend lectures in this branch. He who made us would have been a pitiful bungler if he had made the rules of our moral conduct a science. For one man of science, there are thousands who are not. What would have become of them?"[5] Kant, in fact, was inclined to set the ordinary person's moral sense above that of the philosophers since "it may hope to hit the mark as well as any philosopher may assure himself he will, indeed may here be almost more certain than the latter because he has no other principle than it has; his judgment, however, can be confused and deflected from the right direction by a lot of inappropriate and irrelevant considerations."[6] And here too he was in agreement with Jefferson: "State a moral case to a ploughman and a professor. The former will decide it as well, and often better than the latter, because he has not been led astray by artificial rules."[7]

Kant and Jefferson had similar views of religion. They believed in God, but not in revelation or miracles. They admired Jesus for his moral teachings. For both Kant and Jefferson, Isaac Newton (1642–1727) was a cultural monument. Jefferson counted him among "the three greatest men that ever lived, without any exception."[8] In all his endeavors, with the famous exception of his finances, Jefferson searched for the insight and order that were so magnificently evident in Newton's science.[9] Measuring and experimenting were second nature to Jefferson. "For measuring," Jack McLaughlin reports, "he had his surveying equipment, pocket sextant, theodolite, thermometer, barometer, pedometer, odometer, and a clever wind vane for recording wind directions at Monticello."[10] So was the application of Newtonian mechanism to devices that lightened the burdens of daily life—ingenious clocks, doors, ladders, a turntable clothes closet, lazy susans, dumbwaiters, and more.[11]

Kant wrote his first ambitious work (1755) on cosmology and under the tutelage of Newton, and it was *Carried Out According to Newtonian Principles*.[12] In it, Kant praised Newton for his "most profound insight into the excellence of nature" and "the greatest reverence for the revelation of divine omnipotence."[13] Although the later Kant came to reject all claims about evidence of God in nature, a large part of Kant's mature philosophy can be read as a response to Newton's physics, an attempt to understand its deepest presuppositions.[14]

Discovery and Insight

Newton set the standard for what it is to shed light on reality—to discover and establish order in the world. Kant and Jefferson would have agreed with Alexander Pope's (1688–1744) couplet:

> Nature, and Nature's law lay hid in Night.
> God said, Let Newton be! and all was Light.[15]

Discovering and uncovering were part of the cultural explosion that within half a millennium has changed the face of the earth. It was ignited in sixteenth-century western and southern Europe. There were older cultures at the same time, more refined cultures, perhaps more venerable ones, but none had the restless energy that propelled the voyages of discovery by Christopher Columbus (1451–1506), Ferdinand Magellan (1480–1521), Francis Drake (1540–1596), and many others. These were discoveries for Europeans, of course, not for the indigenous people they discovered. The novel and unique part of these voyages was that for the first time they gave humans a global awareness of the earth.

Discovery not only opened up the world that surrounds us but also the world that is you and me—the human body. Anatomy and dissection had been practiced by other cultures in ancient times, but

Andreas Vesalius (1514–1564) was the first to uncover thoroughly and to present clearly the skeleton, the musculature, the vascular system, the nervous system, the gastrointestinal system, the heart, the lungs, and the brain of human beings.[16] His description of the skeleton very nearly meets today's standards of precision and accuracy.[17] Finally, the exploration of the cosmic world broke through age-old barriers when Galileo Galilei (1564–1642) trained a telescope on the moon and, seeing its mountains and valleys, resolved to build better telescopes and went on to discover four of Jupiter's satellites, the phases of Venus, and more. Yet the most consequential discovery was Galileo's law of falling bodies. It contradicted the view of folk physics that a heavy thing falls faster than a light one. Galileo not only showed that all falling bodies accelerate at the same rate but also that the way things fall can be demonstrated mathematically.

This achievement established modern theory as the standard of insight. If you want to understand something, you have to have a theory of the thing—an explanatory account that centers on laws, principles, or rules. The notion of theory and its explanatory force reached its apogee in the seventeenth century when Isaac Newton published his *Mathematical Principles of Natural Philosophy* (1687).[18] The striking simplicity and generality of Newton's three laws of motion unified and explained the countless ways things move in heaven and on earth. Attempts over thousands of years had failed to specify and relate to one another the forces, distances, and velocities we observe in things when they move. Galileo made headway by casting the lawfulness of terrestrial motion in the language of mathematics, and Johannes Kepler (1571–1630) did the same for celestial motion. When Newton succeeded in stating the laws of any motion whatever, his achievement persuaded many, up into the nineteenth century, that every explanation worthy of its name would have to be mechanical in character, that is, would show that underlying the appearance and behavior of things was the lawfulness of matter and motion.[19]

For Jefferson, Newton's science was an inspiration to seek reasonable solutions for practical matters; for Kant it was a spur to set theories on rational foundations. In part this is a difference between the new world and the old. Jefferson in his younger and mature years was endlessly energetic and optimistic, eager to shape and strengthen the young republic. Kant witnessed the collapse of an old order and the endless pains of a new order struggling to be born. For Kant, to find a rational theory was to discover the abiding structure beneath the unpredictable and confusing features at the surface.

Although lawlike theories set the modern standard of explanation and insight, lawlike explanations (so-called nomological-deductive) are only one species of explanation, the kind that fits physics perfectly, chemistry well, and biology in part. It's a species, moreover, that often crowds out two other, equally important, kinds of explanation. Social theories typically illuminate their subject matter by showing how they are governed by some pattern or paradigm, the market, for example; a certain kind of struggle for power; or a feedback system. The germane kind of explanation in the humanities is a sort of disclosure that points out the salient centers and features of a historical event, a poem, or a painting. We may call these two types of explanation paradigmatic and disclosive, and most of the explaining in the book before you is of those two kinds.

The Charms of Principles

Strictly speaking, Kant did not believe that there could be something like a rationally demonstrable theory of ethics. The force of the moral law, he thought, makes itself felt in our sense of duty, not through rational proofs. Moral reason was practical, not theoretical. Yet in the end, Kant could not escape the hold of theory in a deeper sense—the conviction that a law or principle is at the bottom of moral conduct, that the principle can be uncovered, and that innocent common folk need the help of the philosopher after all. As Kant put it, "[i]t is a splendid thing as far as innocence is concerned, except that it is also terrible that it cannot preserve itself and is easily seduced. Therefore wisdom itself, which otherwise consists more in doing and omitting than in knowing, also needs scholarship, not to learn from it, but to obtain entry and durability for its precepts."[20]

For Kant, the visible and tangible life of moral habits and decisions was a flotilla of storm-tossed ships, very much in need of guidance, and Kant's ambition was not merely to provide landmarks for reflection and orientation, but to furnish a moral compass of rigorous efficacy and to show that human reason, "with this compass in hand, would in all possible cases know how to determine well what was good, what was bad, what was in agreement with duty or against duty . . . "[21] But such ambition overtaxes ethical theory.

In "On the Common Place 'This May Be Right in Theory but Will Not Work in Practice,'" Kant has us consider the quandary of a man who has been entrusted with the safe-keeping of valuables.[22] As Kant has it, the heirs of the late owner cannot possibly know or find out that the valuables really belong to them. The trustee, meanwhile, and his wife and children live in poverty and misery while the heirs are rich, wasteful, and thoughtless. What is the poor man to do? Should he, by returning the trust, try to gain the kind of recognition that will serve him well? If, to the contrary, he uses the deposit to relieve his family's distress, will the reversal of his fortunes arouse suspicion? Yet if he were to proceed cautiously in using what has been entrusted to him, might not his family's misery become irreversible?[23]

In the moral life of ordinary people, Kant argued, the moral law presses its claim through the sense of duty. (Hence the kind of ethics Kant championed is called *deontological,* from Greek *deonta,* meaning *owed.*) Accordingly Kant's unfortunate protagonist will be endlessly worrying and waffling as long as he tries to figure out what is most advantageous for him, but the good man, "if he asks himself what duty requires in this case, will not at all be unsure of the answer he should arrive at, but will be certain on the spot as to what he has to do."[24]

Jefferson was not unfamiliar with the quandaries that arise between the passing of religious doctrine and the uncertain rise of secular ethics, and his advice is similar to Kant's. Writing to Peter Carr, he said: "If ever you find yourself environed with difficulties and perplexing circumstances, out of which you are at a loss how to extricate yourself, do what is right, and be assured that that will extricate you the best out of the worst situations."[25] Jefferson, however, thought we would know "what is right" because we are all "endowed with a sense of right and wrong" and that this sense "is as much part of [our] nature as the sense of hearing, seeing, feeling."[26]

Kant considered the world of the senses unreliable if not treacherous. His foundation of morality is not a sense or feeling but a principle of practical reason "which, to be sure [human reason] does not always think of explicitly in its general form, yet always actually has in mind and uses as the yardstick of its judgment."[27] Of the three versions of the moral law, Kant took the first to be the effective moral device. You are better off, he said, "if in moral judgment you always proceed according to the strict method and use as a basis the universal formula of the categorical imperative: *act according to that maxim that simultaneously can be made into a universal law*."[28]

Quandary Ethics

This then is Kant's picture of how the moral law gains purchase in real life. The typical situation is a quandary, where, as Jefferson put it, "you find yourself environed with difficulties and perplexing circumstances." In such circumstances, a plan of action suggests itself. For the poor man with the trust, it is the idea of taking the valuables. This is his maxim. Now the crucial question is whether the maxim can be universalized without yielding a contradiction or absurdity. So the man asks himself: Can I want everyone to take another's property? Hardly. Stealing makes no sense if I want everyone else to steal as well. It works only if everyone—with my exception—obeys the right of property.

Kant delighted in demonstrating the moral force of the categorical imperative in a variety of cases.[29] But in all the persuasive instances, the quandaries are tests rather than moral crises. In the culture of the poor man, the norms of honesty and property were so well entrenched that there was no genuine alternative to turning the valuables over to their rightful owners. The question was not so much *what* the man should do as to *whether* he would get himself to do the right thing.

Any quandary admits of countless correct descriptions and of as many maxims flowing from them, and it is always possible to find a description and maxim that, when universalized, tells me not to do what I don't want to do.[30] Say the valuables consist of one hundred gold coins. So the poor man frames this maxim: I will take one hundred gold coins to the heirs. Then he universalizes: Everyone must take one hundred gold coins to the heirs. Does that make sense? Are people to mortgage their farms and homesteads to come up with the coins? This is absurd. Hence it is wrong for me to take one hundred gold coins to the heirs. Our moral intuition tells us that this last maxim is silly and that Kant's is correct. But having to rely on intuition in picking the right maxim is fatal to Kant's claim that the *universalizing step,* rather than experience and intuition, reveals which maxims are moral and which are immoral.

There are still deeper problems that surface in quandaries, cases where the situation is not a test of rectitude but a crisis of a person's life, where the question is not whether I am able to square my action with well-established norms, but whether I can discover landmarks that will help me to find my way. Such a crisis befalls Lieutenant Gustl in Arthur Schnitzler's (1862–1931) eponymous story.[31] Gustl is an officer in the Austro-Hungarian army. One fine evening he finds himself attending a concert in Vienna. When it's over and he goes for his coat in the cloakroom, he gets into an impatient tussle with a big man whom, when he turns around, Gustl recognizes as baker Habetswallner. The baker presses his hand on the hilt of Gustl's sword and quietly scolds him and calls him a "stupid boy." Gustl is unable to draw his sword, as the military code requires, and to defend his honor by challenging the baker to a duel. But then the baker relents and with a few pleasantries takes his leave. Bourgeois folk, overpowering and humiliating officers, must have been the bane of the Austro-Hungarian army. In Gottfried Keller's (1819–1890) story "The Prefect of Greifensee," set in the late eighteenth century, Marianne, who works as a cook in a Freiburg im Breisgau inn, feels slandered by a jealous Austrian officer. Armed with a kitchen knife, she confronts the officer in the dining hall; when he draws his sword to deal with her, she disarms him, breaks his sword, and throws the pieces at his feet, whereupon the man is expelled from his regiment.[32]

This is the fate that stares Gustl in the face, and he spends a sleepless night, wandering about Vienna and wondering what he should do. Evidently his problem is not whether he can live up to an obvious maxim, but what maxim or option he should choose among the ones that come to his mind: (1) Hoping that the baker will keep silent about the incident, Gustl could pretend nothing has happened. (2) Gustl could report the incident to his superior and hope for understanding. (3) He could take his case to the military court of honor and ask for a favorable judgment. (4) He could shamefully quit the army. (5) He could emigrate to America. (6) Or he could shoot himself. Gustl considers all options, the last one most seriously.

Yet the dry enumeration of ways to proceed does not do justice to the pitiful turmoil of Gustl's soul, where the norms of modern ethics are rising in the background but are too faint to help him. Why this inequality in the burdens of honor between burghers and the officers? Should not Gustl's dignity prevail over the strictures of an inhumane military code? What does self-determination demand of Gustl?

Schnitzler gives us a portrait of a personal crisis, of a man whose identity and intentions have been upset by a trivial incident. What makes the crisis personal is the apparently unyielding solidity of the social conventions that are surrounding Gustl. But couldn't Gustl step out of this confinement, emigrate to America, and leave the Old World to the social crisis that was impending with the First World War? Not that the New World has been entirely free of the senseless honor code. Admiral Jeremy Boorda, it appeared, had not quite deservedly been wearing a combat medal, and when in 1996 this was about to become public knowledge, Boorda shot himself.

Gustl does in fact think of emigrating to America. He knows it is a refuge and a place for new beginnings. But he feels unequal to the challenge.[33] He is incapable of a fundamental change. When the new day dawns and Gustl has breakfast in a coffee house, he learns that baker Habetswallner died of a heart attack on his way from the concert. Gustl shakes off his soul searching and continues in his supercilious ways.

The moral of the story is that the landmarks of theoretical ethics give us orientation that is as broad as it is indispensable. They give us a rough bearing, but they leave the regions of daily life unmapped, and their very visibility can become blurred in a moral crisis, personal or social. Kant's attempt to make theory not just necessary but sufficient for ethics and to turn a landmark into a device, had to fail.

CHAPTER SIX

The Pursuit of Happiness

Jefferson and Mill

From the beginning of the republic, equality and prosperity have been competing with each other. In the early years, and broadly speaking, the Republicans championed equality, the Federalists prosperity. Jefferson was the most prominent Republican, and he thought that equality and liberty would best be served in a decentralized and agrarian society. In his later years, Jefferson accepted the necessity of manufacturing and industry.[1] But the growing power of the Federal Government—a necessity for the vigorous advancement of prosperity—filled his old age with gloom as the Enlightenment in American culture was being eclipsed by industry and commerce. The shift has been memorably chronicled by Henry Adams (1838–1918), the great-grandson of Jefferson's sometime enemy and eventual friend and correspondent, John Adams (1735–1826). Henry Adams thought of himself as "an eighteenth-century child," born in the nineteenth century, and never in tune with "its money-lenders, its bank directors, and its railway magnates."[2] The movement that displaced reason and civic virtue had its own theory.[3] John Stuart Mill (1806–1873) gave it its classical statement in *Utilitarianism*.[4] Henry Adams had in fact met Mill and thought of him as fey and mistaken.[5] Karl Marx, Adams thought, was the thinker to be reckoned with.[6] After a century's struggle between the followers of Mill and Marx, Mill's view, or at least a version of it, seems triumphant and without rival unless a version of Jefferson's vision is reborn.[7]

The huge obstacle that lies in the path toward a renewal of Jefferson's vision is the apparent agreement of Jefferson and Mill. They agreed that life should be devoted to the pursuit of happiness. It looks, therefore, as if Mill was the heir and executor of Jefferson rather than his opponent and vanquisher. Jefferson memorably included "the pursuit of happiness" among the "unalienable rights" of all men, and utilitarianism, as Mill understood it, was based on the "greatest happiness principle."[8] But did Jefferson and Mill mean the same thing by "happiness"?

We can infer what Jefferson meant from his times, his readings, and his writings. The pursuit of happiness he took first of all to be a natural human inclination that sought its development and flourishing in this world rather than in a future world. Next, education and knowledge were crucial for the successful pursuit of happiness. Finally, an individual's happiness was inseparable from the happiness of the community. Individuals could promote their happiness only by promoting the happiness of all.[9] As a consequence of all this, happiness is impossible, Jefferson wrote, in a setting "where ignorance, superstition, poverty and oppression of body and mind in every form, are so firmly settled on the mass of the people, that their redemption from them can never be hoped."[10]

The United States was to be the country where these evils would be removed and where conditions would allow one to pursue happiness. Jefferson said little, however, about the actual shape of happiness. With allowances made for the injustices of his time, Jefferson answered the question, not in the writings he left but in the life he lived. Mill to the contrary gave an answer that seemed explicit and convincing. "By happiness," he said, "is intended pleasure and absence of pain."[11] Mill was worried that his readers might understand "pleasure" in too crude a sense, and he tried to argue that one had to assign "to the pleasures of the intellect, of the feelings and imagination, and of the moral sentiments a much higher value as pleasures than to those of mere sensation." "Mere sensation" was a euphemism for food and sex, the pleasures "worthy only of swine."[12]

But this is mere pleading. The Mill who wrote *On Liberty* (1859), and the great utilitarians, all agreed that the individual is to be the authority on what a person finds to be pleasing and pleasurable.[13] Thus utilitarianism turned into an affirmative, tolerant, and seemingly democratic social philosophy. It appeared to take the pursuit of happiness seriously. "The greatest happiness for the greatest number of people" was framed as the principle: Maximize happiness for an entire population.

Utilitarianism was no less charmed by theory and its promise of simplicity and sufficiency than Kantian theory. Anyone who knows the details and development of utilitarianism will smile at the suggestion that it has a charming simplicity. The endeavors to make it work have led to endless and forbidding complexity, granted. But it is the inaugural straightforwardness of the theory that has inspired and has again and again refreshed the energy of economists and philosophers.[14]

The Dark Sides of Utilitarianism

Yet the technical problems that beset utilitarianism are more than matched by its moral problems. Each of the three elements of the utilitarian principle—maximizing, happiness, and regard for an entire population—has a dark and dubious side. The devotion to maximizing has led to a frenetic, exhausting, and sometimes to a crushing style of life among the country's elite; even the rest of us, who work fewer than sixty hours a week and see our loved ones every evening, are made to feel like slackers and failures.[15] Most troubling perhaps is the slowly dawning realization that all this pushing, struggling, and pursuing is not getting us any closer to happiness.

For happiness to be maximized, it needs to be measurable in some kind of currency, otherwise the devotion to "the greatest happiness" is just hand waving.[16] Given such a measure, we are able to weigh satisfaction against dissatisfaction and aim at a net maximum of pleasure. The first problem with measurement is searching for an objective quantitative standard. To find one is a complex task at best. But let us assume we have found one. Then a yet-deeper problem surfaces. Such a measure would distort the moral life by reducing the great variety of good things to one kind of value and allowing us (forcing us—if we stick to the principle) to trade off one against the other as maximizing dictates. Bernard Williams has illustrated the problem: "If racist prejudice is directed toward a small minority by a majority that gets enough satisfaction from it, it could begin to be touch and go whether racism might not be justified. The point is not how likely that is to arise, or in what circumstances, but that the whole question of how many racists are involved cannot begin to be an acceptable consideration on the question whether racism is acceptable."[17]

The illustration also reveals the calamity that lurks beneath the apparently democratic principle of taking everyone's satisfaction into account. As utilitarians we must count the negative value of a persecuted minority's pain, but we must also compute the net of the minority's misery and the racists' pleasure, and if that net is greater than the satisfaction of a secure minority minus the displeasure of racists, we must go after the net maximum. More generally, if the mistreatment of some part of the population leads to an overall increase of satisfaction, the utilitarian ought to mistreat those people.

When it comes to justice and compassion, utilitarianism shows a personality split between generosity and selfishness. The generous side has been well represented, in theory and practice, by Peter Singer. He has often pointed out that, when we have discretionary funds to spend, utilitarianism requires us so to spend them so that satisfaction is maximized. Say I have $100,000 to spend, and I suffer from a midlife crisis. To cure it, I am thinking of buying a BMW sports car. It will give me satisfaction—heads will turn, the wind will rush about me, twisting mountain roads will become prey of my prowess.

As I am deciding on the color and extras, my utilitarian conscience prompts me to ask: Will the money, spent on this car, produce the greatest happiness? What if I buy four Toyota station wagons and give them to single mothers without wheels, women who struggle to get the groceries home and who can't think of taking their children out into the mountains? Mill was familiar with a more general and genteel version of this problem and knew that some people found utilitarianism too demanding because "it is exacting too much to require that people shall always act from the inducement of promoting the general interests of society."[18] But Mill reassures our troubled conscience: "The multiplication of happiness is, according to the utilitarian ethics, the object of virtue: the occasions on which any person (except one in a thousand) has it in his power to do this on an extended scale—in other words, to be a public benefactor—are but exceptional; and on these occasions alone is he called on to consider public utility; in every other case, private utility, the interest or happiness of some few persons, is all he has to attend to."[19]

Mill's characteristic moderation and imprecision make the point hard to argue with, but we are inclined to hear in those remarks the echo of more pointed sayings that had come from Adam Smith (1723–1790) more than a century earlier: "Every individual is continually exerting himself to find out the most advantageous employment for whatever capital he can command. It is his own advantage, indeed, and not that of the society, which he has in view. But the study of his own advantage, naturally, or rather necessarily, leads him to prefer that employment which is most advantageous to the society."[20]

It's to my own advantage to buy the BMW. Should I? "I am saying that you shouldn't buy that new car," says Peter Singer, "take that cruise, redecorate the house or get that pricey new suit. After all, a $1,000 suit could save five children's lives."[21] You should maximize happiness for an entire population. For Singer, the relevant population, to a first approximation, is the human community, and surely five children saved from dying feel more satisfaction than I do wearing the suit or driving the car.[22]

Singer is right on this point. But utilitarianism has this in common with most ethical theories: It works well with clear and urgent cases, or more accurately, it is not needed for such cases. We know in dire circumstances without the help of theory what is right even if we can't get ourselves to do it. An ethical theory is put to its real test when it is applied as a comprehensive view to the subtle and complex issues of today's typical human condition.

Assume Singer's appeal is "exacting too much" at this time, and take as the relevant population the citizens of this country rather than the global population. What then needs to be done to obtain a maximum of satisfaction? There is no answer to this question unless, to come back to the earlier point, we have an answer to a prior question: Is there a valid and reliable unit of measurement (a so-called cardinal measure) for utility, happiness, or satisfaction? Two answers have been given to this question. The first gives up on wanting to look into people's heads to see how happy they feel. Instead, it simply looks at the observable behavior of people, at the decisions they make when spending their money. Those decisions reveal their preferences, and those can be ranked and modeled ordinally, that is, according to what people prefer to what, regardless of why or how much they prefer this to that.

Monetary Utilitarianism

There are two problems with this answer. The first is that people regularly make choices that are counterproductive to the happiness they want. The second is that money slips in as a unit of measurement, and the sum of all goods and services, expressible in money, that is, the Gross National or Domestic Product (GNP or GDP) becomes the gauge of satisfaction.[23] But why don't we simply accept what the economy suggests and everyone believes—money is the measure of happiness, and the relevant population is our society? If we do, we get what can be called monetary utilitarianism—the second answer to the measurement problem.[24]

For all its psychological and moral flaws, this is a powerful and revealing version of the theory. It is powerful because it lends sharp contours to the defining terms of utilitarianism. Money excels both as a motor and a measure of maximizing. The common fervor of the search for fortune needs no proof of existence. As a measure, money lets us get a grip on the sum of all the stuff that is supposed to make us happy, the goods and services annually produced within this country—the Gross Domestic Product (GDP). It allows us to measure yields as interest, dividends, and appreciation. It helps us to measure productivity and is congenial to associated nonmonetary measures such as the unemployment rate and consumer confidence. We are endlessly fascinated with these measures and worried about them.

Money has a dubious reputation as a measure of satisfaction. We all claim to know that money can't buy me love nor can it buy me happiness. But this is partly wrong. Affluence and happiness rise together up to a point, both among and within societies. Moreover, we don't quite believe what we claim to know. Yes, the rich are no happier than I am, and, yes, after the first exuberance the lottery winner's happiness declines to what it was before. But I know I could pull it off—I could be very rich and enduringly happy both. The belief that my next self-indulgent purchase will bring me greater and lasting happiness is as unshakable as it is mistaken, and it is reinforced when my nose is rubbed in relative poverty. The journalists who report on Hollywood stars are surely comfortable economically, and in Washington, Bernard Weintraub has said, they "do not feel diminished by their lower salaries. In Hollywood, many do. I did. . . . For many of us on the press side, the money gap leads to resentment and envy."[25]

Yet these sentiments are as unwarranted as they are plausible. Once we, as individuals, have reached a moderate level of affluence, additional prosperity produces no more happiness, and as a society, we have long passed the stage where more affluence could make us happier. Happiness, if anything, has been declining in this country.[26]

Monetary utilitarianism is more illuminating still when we restrict the relevant population to that of the United States because we then get a handle on who has how much and whether the sum of it all has been moving up. But it also deals with a problem that Bernard Williams, with typical wit, has described this way: "Utilitarian benevolence involves no particular attachments, and it is immune to the inverse square law."[27] This is also the virtue of utilitarians. For Peter Singer, all people, in theory, have the same claim on his benevolence, be they near or far. For the rest of us, concern diminishes with the square of the distance—very rapidly. But there is something unnatural about utilitarian even-mindedness. It seems right, if theoretically hard to argue, that my wife and daughters have a stronger hold on my affection and support than my neighbors, and my neighbors in Missoula a stronger claim than my fellow citizens in Montana, and so on. Singer himself has not in fact been immune to the inverse square law. He has spent disproportionate amounts of money and care on his mother, rightly in my view, but in violation of his utilitarian principles.[28] At any rate, focusing our attention on our country is what we should do first though that cannot be our exclusive concern.

Monetary utilitarianism is revealing because it shows clearly how liable to injustice utilitarianism is and because it begins to show how the current notion of happiness distorts and impoverishes our lives. The moral Achilles heel of utilitarianism is its commitment to the pursuit of overall happiness no matter what happens to the happiness of some individuals. In monetary utilitarianism this amounts to pursuing overall prosperity even if some people fail to share in the growing affluence, and monetary utilitarianism allows us to make this crisp and clear. Prosperity is the annual sum of all goods and services (the GDP) per person in the United States, also called the standard of living. Since 1979 it has grown (in 1996 dollars) from $21,821 to $32,839 in 2002.[29]

There are ways of making inequality painfully clear—you quantify it. One quantification method, called the Gini coefficient, gives you an overall measure of inequality, a number between 0 (total equality) and 100 (total inequality). Both total equality and total inequality are immoral and unfeasible. Hence what you find among the industrialized and high-income countries is a number somewhere between 30 and 50. In the United States it has risen from 39.4 in 1970 to 45.6 in 1994.[30] Internationally, the United States was fifth worst in the order of equality among twenty-one comparable countries. Our Gini coefficient was 35.28, the United Kingdom's 25.98.[31]

Another way of measuring inequality is to compare what the bottom fifth (or tenth) and top fifth (or tenth) make. From 1979 to 2001, the family income of the bottom fifth in this country has grown by 3 percent, while that of the top fifth has risen by 53 percent. Again in the 1990s, the ratio of household income of the 10 richest to the 10 poorest percent in this country was by far the worst among the seventeen industrialized democracies. Ours was 5.44, Sweden's was 2.59.[32] To sum it up, while American prosperity has grown significantly in the last generation, so has inequality, and our inequality is worst among comparable countries. At the same time, mobility has decreased—more of the poor stay poor and more of the rich stay rich.[33] Poverty, moreover, is no longer simply having less of whatever the others have. In important regards, poverty is being without some of the basics that would help you to get out of misery—no health care, no functional education, no day care, no access to a decent job, no effective legal advice, no access to the Internet.

One reason why we as a society have become so indifferent to our responsibility for the poor is the continuing transformation of American culture that is also revealed, if more tentatively, by monetary utilitarianism. Money as the decisive prod and measure of happiness has had a leveling and distorting effect on the depth and richness of life.[34] The late twentieth-century offspring of Marx have coined a term for this process of distortion. They call it commodification. It carries connotations of disapproval unlike the term that conservatives prefer—privatization—or the term of mixed connotations—commercialization.

Commodification

Commodification has a clean and crisp economic definition—it is the process of moving something into the market so that it becomes available as a commodity, that is, for sale and purchase. Moved into the market from where? From the intimate sphere or the public sphere. In the latter case, a public good is converted into a commodity, and, speaking more precisely, privatization is commodification in this sense only. Some of the public goods, such as justice and elementary education, are not material; others, such as transportation or a healthy environment, clearly are. The same distinction applies to intimate goods. Friendship and a sense of belonging are not material goods, but food and clothing are.

Commodification of some intangible goods is morally objectionable because here a good commodified becomes a good corrupted. Justice bought is no longer justice, and friendship paid for is not real friendship. But no such opprobrium seems to taint tangible goods. Railroads used to be managed as public goods by European governments. In the United States, they are for the most part private enterprises run for profit. Food and clothing have left the intimate sphere of the household so long ago that we no longer notice their peculiarities as commodities. Michael Walzer, who has thought deeply about commodification (though he does not use the word), has drawn up a list of never-to-be-commodified goods, all of which, not surprisingly, are intangible.[35] The way commodification corrupts justice and friendship is ethically obvious. But even if we prevent such commodification, the system of which it is a part, monetary utilitarianism, remains a moral threat to public and intimate goods. Monetary utilitarianism is bent on maximizing, and it maximizes what is maximizable—things in the market, that is, commodities. The rest is neglected. Advocates for Chevies and Fords are plentiful and take long coffee breaks. Advocates of justice for the poor (public defenders) are in short supply and terribly overworked.

To make these moral observations about commodification, you can rely on its economic definition. The ethical problems are boundary issues: What is in? What is out? What happens to what is out? What has become of what is in? Yet some moral qualms we have with commodification touch the very nature of commodification, or better, touch the cultural and moral, rather than the economic, character of it. Consider surrogacy. The fifth item on Walzer's never-to-be-commodified goods is procreation (along with marriage). But it is in fact routinely commodified through the purchase of surrogate motherhood. Richard Posner gives us a disarmingly simple argument in favor of enforceable surrogacy contracts:

> Such contracts would not be made unless the parties to them believed that surrogacy would be mutually beneficial. Suppose the contract requires the father and his wife to pay the surrogate mother $10,000 (apparently this is the most common price in contracts of surrogate motherhood [in 1989]). The father and wife must believe that they will derive a benefit from having the baby that is greater than $10,000, or else they would not sign the contract. The surrogate must believe that she will derive a benefit from the $10,000 (more precisely, from what she will use the money for) that is greater than the cost to her of being pregnant and giving birth and then surrendering the baby.[36]

This is a small-scale triumph of monetary utilitarianism. Total satisfaction seems to be maximized with money serving as the measure of value.

Some of us will find such commodification distasteful or immoral, but Posner likes it, and he has an argument to support his preference. Do opponents have arguments? They claim surrogacy is baby buying. Defenders say it is fee for service. Opponents say it exploits poor and less-educated women. Defenders say surrogate mothers are free and autonomous women. It looks like a standoff. To get beyond it, we need not just an economic but also a moral conception of commodification, and before that we may need a better understanding of happiness.

Measuring Happiness

Perhaps, then, we should revisit the questions: What is happiness? And how do you measure it? To start with, we may be inclined to agree with Aristotle (384–322 BCE), who begins by noting that all kinds of people pursue happiness and then says:

> But what constitutes happiness is a matter of dispute; and the popular account of it is not the same as that given by the philosophers. Ordinary people identify it with some obvious and visible good, such as pleasure or wealth or honour—some say one thing and some another, indeed very often the same person says different things at different times: when he falls sick he thinks health is happiness, when he is poor, wealth. At other times, feeling conscious of their own ignorance, people admire those who propound something grand and above their heads; and it has been held by some thinkers that beside the many good things we have mentioned, there exists another good, that is good in itself, and stands to all those goods as the cause of their being good.[37]

This sort of disagreement is aggravated by contemporary subjectivism and relativism. Happiness is whatever. For Aristotle it is a life of virtue; this is the goal or good that should govern our conduct. Goal (or distance) in Greek is *telos*. Hence Aristotelian ethics is called *teleological*. But the good or goal, as Aristotle acknowledges, can be variously defined, and Mill's ethics of pleasure is teleological as well. Teleological ethics, that is, the pursuit of the *good,* whatever its definition, contrasts with deontological ethics—doing what is right, come what may.

In the last thirty years, the apparent elusiveness of happiness has been limited by social scientists who have found that happiness can be validly described and reliably measured. Techniques for measuring happiness or well-being are still being refined, and there is a need for such refinement because people are subject to systematic mistakes in assessing their happiness. Looking back, people fail to sum their positive and negative emotions accurately. When it comes to pain, for example, they give undue emphasis to the peak and to the conclusion of a painful episode. If at the conclusion pain was relatively mild, people remember the event as mildly painful even if it was consistently quite painful up to its conclusion. Looking ahead, people overestimate the force and the duration of the pleasure they are planning or anticipating, and looking at the satisfaction of their lives as a whole, they judge it as much by social expectations as by actual experience.

A recently designed method, the Day Reconstruction Method (DRM), is intended to reduce "the errors and biases of recall" and the effects of "belief-based generic judgments," that remind people of "instances that are prototypical but not necessarily typical" and that discourage "reports of socially inappropriate affect."[38] In retrospect, people, mindful of what society expects of them, report that taking care of their children is more enjoyable than watching television.[39] But a DRM study of 909 working mothers in Texas found the opposite to be the case.

If we think of happiness as the net of the pains and pleasures we actually experience, then it is plausible to think that the errors and distortions of recall and anticipation effect a reduction of the amount of happiness individuals could experience. They should increase those episodes that the DRM and similarly objective methods have shown to be actually pleasurable or more pleasurable than other episodes.

There appear to be equally important social consequences of refined measurements of happiness. The article that introduces and illustrates the DRM ends with this sentence: "The DRM or its variants could also contribute to the development of an accounting system for the well-being of society, a potentially important tool for social policy."[40] One commentator drew an explicit analogy of such an accounting system to the Gross Domestic Product. Richard Suzman of the National Institute of Aging said of the DRM: "At the broadest level, it could help us set up a national well-being account, similar to the gross national product, that would give us a better understanding of how changes in policy, or social trends, affect the quality of life."[41]

An objective and accurate National Well-Being Account could replace monetary utilitarianism with a direct and clean version of utilitarianism. There is no doubt that happiness research in all its forms and especially in its refined and sophisticated models is of great social and moral significance. If it turns out that we as a society make humanly exhausting and environmentally harmful and yet largely futile efforts to raise overall happiness, then we better think about what we are doing. At the same time, the answer to the question "What now?" is not the utilitarian principle "Raise the National Well-Being Account to its highest possible level." Profound practical and moral problems stand in the way.

Technically, the scale that is used to measure feelings of happiness compares poorly with money. Although certain public and intimate goods are systematically excluded from measurements in dollars, an infinite number of things have a price. A first-class stamp has a price, and so does the national debt of the United States. The emotional life for its part has a plethora of kinds and contours of feelings, but there is little chance that all of them will get a measurement.

Money, moreover, gives us fine-grained measurements, from the cents of a stamp to the trillions of dollars of the national debt. Finance, along with computer science and astrophysics, has taught us to use scales of a dozen or more orders of magnitude. By contrast, the scale used to measure happiness in the DRM study allows for roughly six hundred degrees of difference.

But even six hundred degrees of pleasure may give us a false or loopy sort of precision. Money allows us to make comparisons and decisions we understand well. If in a particular year we have lost $20,000 in the stock market and come into an inheritance of $40,000, we know that at the end of the year we are better off. We also know that with the additional $20,000 we could buy a new car or remodel the kitchen or pay a year's college tuition for one of our children. But say half an hour of intimate relations is rated 5.10, and half an hour with the boss is 3.52, yielding an average of 4.31 for the hour. As it happens, exercising is rated 4.31. Should we conclude that, other things being equal, the day of intimate relations and the meeting with the boss yielded as much happiness as the day when we exercised for an hour? The "accounting system for the well-being of society" then compares unfavorably in scope, resolution, and fungibility with monetary accounting. Of course the two kinds of accounting are not necessarily rivals. They could exist side by side and be useful together.

The Shape of Happiness

From the moral point of view, the well-being account needs to be hedged about with ethical norms and reflections even more so than monetary accounting. The life of pleasure is not necessarily the good life, and a society whose well-being account has been rising could be in the process of becoming a smug and self-centered society. The crucial question in all of this is: What is the ethically authoritative point of view?

If the women in the DRM study had been engaged in conversations about their lives and in the course of the conversations had been asked what they enjoyed more, taking care of their children or watching television, they probably would have said that taking care of their children was more enjoyable.[42] But in light of the DRM findings, that answer would have reflected "distortions" and "errors and biases" because it would have been based on "belief-based generic judgments ('I enjoy my kids')" rather than "specific and episodic reports ('but they were a pain last night')."[43] On a 0 (not at all enjoyable) to 6 (very much enjoyable) scale, the women ranked "watching TV" 4.19 and "taking care of my children" 3.86.

What conclusion should we draw from this? That to increase their well-being, women should watch more television and reduce the time it takes to look after their children? Presumably, if the women had been asked, episodically or generically, which is more important or valuable or worthwhile, television watching or child care, they would have said that "Taking care of my children" is more important and meaningful than "Watching TV."

Assuming that to have been the answer, we can see how empirical findings and ethical considerations bear on each other. There is a difference between an ethical norm and a score on a scale of pleasure. "It is a measure of people's mood in the moment," the lead author of the DRM study, Daniel Kahneman said, "but that does not mean that it's the best thing they could be doing."[44] There is, at the same time, an important tie between empirical findings and moral standards. If what people consider worthwhile they also find distasteful, something cries out for reflection and reform. In the case of working women who carry the major or the entire burden of child care, it is more than understandable that they feel harried, what with the children's homework, their fighting, their fevers, their dinner, the next day's lunches, the music lessons, the soccer games, and more. A quiet hour of television is rest and relief in comparison. Studies of television, using a method that is even more "objective" than the DRM, have found that television watching is disengaging and becomes depressing as a person watches it for longer periods.[45] Evidently, an hour of television that serves as a respite from a stressful day feels different from three hours of watching at the conclusion of an uneventful day.

One lesson we should draw from this is that we should arrange our lives—the lives of working mothers in this case—so that to the extent possible what is worthwhile is enjoyable too. A second lesson may be that not only do happiness measurements mean little when we compare different kinds of episodes, but that apparently what looks like one kind of episode, watching television for example, feels different depending on context and duration. That suggests a larger lesson. Happiness is not a pile of episodes, separately weighed and summed together in some way. Happiness is found in the shape of a day, of a year, or of a life.

Happiness takes the form of a story that comes to a good conclusion, a story of many subplots and intermediate summaries. Some religious people conclude the day with the compline, a prayer in conclusion as the name suggests. Martin Seligman considers the shape of his life at the conclusion of a year.[46] Aristotle thinks that the real story of whether a person can be called happy or not is told at the end of a person's life, and the fate of one's descendants add codas that make some contribution to the ancestor's happiness.[47] What gives a life its happiness are not the pleasures of good fortune, Aristotle adds, "for fortune does not determine whether we fare well or ill, but it is, as we said, merely an addition to human life; practices in conformity with virtue constitute happiness, and the opposite activities constitute its opposite."[48]

Pleasure comes into its own when it is in tune with virtue and reality, and whether this yields a harmony of happiness depends on the resolution of the chords of life. If it is true that in looking back we can't help but pay disproportionate attention to the conclusion of an episode, there must be an evolutionary advantage in this. "Without it," a student and a mother once said in my class, "no woman would have a second child." Evolutionary dispositions, as we will see, provide important background conditions for ethics. In this case, evolutionary benefit is raised to a moral conclusion—life is a story with a moral, and all's well that ends well.

CHAPTER SEVEN

Evolutionary Psychology

Conservatives, Liberals, Nature, and Nurture

The pursuit of prosperity is the most powerful, social, and political force in this country. It recruits even those to its cause who have fared poorly in it.[1] Quite often there are enough voters among the poor and in the lower middle class who would rather dream of distant prosperity than benefit from greater equality today to hand conservatives the presidency or control of Congress.

Although they are the most trusted promoters of prosperity, conservatives have in fact a distinctive concern—their devotion to tradition—that should make them suspicious of growing prosperity as we know it. As traditionalists, they cherish the heroic figures and lessons of history. They are devoted to religion and family. They appreciate the classic achievements of the arts. From all this they draw an ethics of aspiration and excellence.

Most of the conservative treasures and traditions were gathered and have been maintained by prosperous families. Hence traditional conservatives have always been concerned to preserve and augment their prosperity, and aspiring conservatives have been concerned about amassing a fortune as the basis of their descendants' privileges and graces.[2] Through the Industrial Revolution, the basis of power had shifted from land to business and industry. Ever since, conservatives have been forced to favor technology as the wellspring of their prosperous way of life. Modern technology, however, subverts, exploits, and guts traditional customs and excellence—something George Will rarely lets his fellow conservatives forget.[3]

Lately, to make things worse, information technology has threatened to throw up an entirely new breed of affluent elite that refuses to divide into the two reassuringly familiar branches of prior arrivistes—the punctilious students of gentility and the hopelessly crude nouveaux riches.[4] Although the specter of hordes of affluent geeks has become less ominous, the feeling of power among the conservatives remains heavily tinged with misgivings.

While the conservatives have been troubled by technology, the liberals have been betrayed by science. In 1909 Herbert Croly urged in *The Promise of American Life* that expert knowledge must be used to promote the common welfare.[5] Liberals have since looked to science as a warrant and instrument of their reform proposals. The seeds of trouble were sown with the assumption that human minds are essentially blank tablets from which, if necessary, to erase false and injurious doctrines and on which to inscribe benign and egalitarian convictions. This assumption reached a high-water mark toward the end of the last century when liberal intellectuals considered most everything a social construction and hence subject to deconstruction and reconstruction. Power, gender, ethnic identity, social organization, the sciences themselves, and even the wilderness were said to be socially constructed. All liberals, at any rate, tend to favor nurture over nature as the fundamental factor in the moral shaping of society.

Conservatives, to the contrary, prefer the picture of a stable reality, and they like to think of their typically superior economic position as the outcome of their innate excellence. High achievers of the lower classes are welcome as sons-in-law and vivacious beauties as daughters-in-law, but wholesale reform, conservatives believe, violates the nature of things, and attempts at reform can only lead to mischief or worse. Conservatives like to believe that human nature rather than social nurture is the appropriate basis of the social order.

At the same time, evolutionary biologists, emboldened by the refinements of evolutionary theory through genetics and microbiology, began to extend evolutionary theory to the explanation of human behavior. Since the human body is shaped by evolution, it is plausible to expect the human mind and human behavior to have been molded the same way. There was enough distance to the earlier and unsavory applications of evolution to human behavior—social Darwinism, eugenics, and racism—to make a new beginning. E. O. Wilson made that start under the heading of sociobiology with the publication of the eponymous *Sociobiology* in 1975.[6] When the book was greeted with much hostility (undeserved, I think), Wilson's successors let discretion be the better part of valor and renamed their enterprise "evolutionary psychology."

Evolutionary biologists, however, adhere to the robust realism that most scientists see forced upon them by the nature of their enterprise. If you set out to discover what in fact is the case, you will say (after Bertrand Russell) that social constructivism "has many advantages; they are the same as the advantages of theft over honest toil."[7] A hunger for reality among certain social theorists and the sheer plausibility of the program conspired to make evolutionary psychology a widely and ably propounded theory. Being a theory of human behavior, evolutionary psychology inevitably came to bear on moral conduct.[8]

Sex, Selflessness, and the Ancestral Environment

Sexual behavior and altruism loom large in evolutionary psychology, the former because it is so central to evolution and life, the latter because it seems to count against the logic of evolution. That logic, briefly, pivots on variation, selection, and continuity. Without any one of the three elements, evolution would have been impossible.

Variation of the genetic blueprints of organisms comes about through cosmic radiation, chemical influences, or errors in transcribing one DNA strand into another. Without variation, nothing would change, far less evolve. Selection is enforced by the rigors of the environment, and the challenges that an organism meets or succumbs to importantly include fellow organisms. Without selection, no excellence—every variation would survive and beget more variations, no matter how ugly, awkward, or bizarre the result. But awkward compared with what? Elegance is born of challenges that are simply and gracefully met. Continuity, finally, assures that changes do not get lost, but are preserved from parent to offspring and provide the basis for further changes. Without continuity changes would remain inconsequential episodes.

Everything in the realm of life has come about through the causal mechanism of these three interlocking components of evolution. Of course the actual paths of development are intricate and full of surprising twists and turns. But as in the case of utilitarianism, the endless complexities of evolution appear to resolve themselves in the attractive simplicity of its fundamental law. Evolutionary theory holds the promise of its deontological and utilitarian rivals—to furnish lucid explanations of forbiddingly difficult phenomena, among them the thorny issues of moral dispositions and behaviors.

Consider sexuality. For many organisms, the mechanism of continuity is sexual reproduction. Those organisms in sexually reproducing species that compared with their competitors are unconcerned or inept with sex will soon disappear without a trace. In humans and other mammals the concern with reproduction is typically different for females and males. For a woman, reproductive events are relatively rare and always burdensome; each represents a risk to health or survival. Normally a woman can engage in reproduction twelve or so times. Given the risks, burdens, and scarcity of her reproductive events, female caution and discretion in choosing a mate must have been favored by evolution. Of a man, to the contrary, reproductive occasions require little, and they can be numerous. Hence male sexuality must have evolved to be expansive and aggressive. That division of strategies is detectable among humans.

Yet human sexuality is hedged with complications. Who, in the evolutionary past, would have had more offspring, the philandering man who left his countless children without support or the faithful mate who saw to the welfare of his relatively few children? Did the woman who carefully chose a man and matched his fidelity with hers have more resources for her children than one who had sex with several men and was supported by many of them on the assumption of each that he was the father? And could not a woman have had the best of both worlds by deceiving her faithful but unspectacular mate and having sex with a prodigious and roving man? In that case women who were good at deception would have been most successful. But reliable and unsuspecting men would then have died out unless they became good at detection. Would not this have led to a costly arms race of deceiving vs. detecting capacities? And would not the sexually warring strain of humans come to be at a disadvantage in relation to a more trusting and straightforward strain?

Selflessness or altruism is the dare to evolutionary theory that its proponents were eager to meet. The challenge seems formidable, for how could selfless devotion to the welfare of other people be squared with the struggle for survival? But what really struggles to prevail and multiply, so goes one evolutionary reply, is not the organism or the species but rather genes for whom organisms are merely the vehicles of their competitions. Thus an animal that is genetically wired to sacrifice itself for the sake of five or six of its next of kin will serve the genes of its family better than the one that saves itself and loses several of its relatives to the coyotes. Such altruism, based on kin selection, can clear the way for the more inclusive reciprocal altruism if it turns out that a group of cooperative animals or humans does better than its selfish or merely kin-altruistic competitors.

Sex and altruism reveal a social structure that is deeply rooted in the particular circumstances of evolution. Still more important is the resonance the human condition shows even today to the social and physical environment—the ancestral environment or the environment of evolutionary adaptation—that shaped humans, body and mind. It was a world that was typically limited and intimately familiar, inhabited by a small group of well-known and well-regarded relatives and neighbors, ordered by focal points and landmarks, touched, handled, and animated by the daily practices that sustain and illuminate life. It was also a world of hardships, dangers, and challenges that as often as not darkened or destroyed life. This is the world we have evolved to be familiar and to cope with.

Evolution and Ethics

For the liberal imagination, these evolutionary facts are obstacles to reform. It is inconvenient to be told that the difference in sexual inclinations of women and men is not just an artifact of patriarchy, that the spirit of cooperation is not simply the product of a social contract, and that the circumstances of human flourishing cannot be left entirely to radical innovation and rational design.

That all these facts are of a piece with common sense realism and thus defy radical deconstruction make these findings less palatable still. Postmodern theorists would not, however, have been deterred in their deconstructive zeal had they been in a position to pin the unwelcome news on a conservative cabal. But proponents of evolutionary psychology like Robert Wright, Steven Pinker, Daniel Dennett, and Peter Singer have impeccable liberal credentials. As for the authority of science, Alan Sokal's well-published faux-constructivist article and his revelation of the hoax have taught theorists of deconstruction that ignorance and contempt of the sciences exact too high a price in public credibility.[9]

Yet the orthodox liberal dream of having ultimate and universal jurisdiction over reality, based on the claim that whatever is real to you and me owes its reality to a social construction, subject to deconstruction and reconstruction, was not easy to give up. Thus a break has lately divided the liberals into one camp that preaches the radical openness of the human condition and another that stresses the natural facts of human being.[10]

How tight, in any event, are the constraints that evolution imposes on ethical theory? Can evolutionary theory claim to replace ethical theory? Daniel Dennett has sketched a "moral first aid manual," based on evolution. It contemplates a quandary that meets with a great variety of suggestions and considerations, selects among them by means of "conversation stoppers," and ends with the continued existence of those solutions that prevail in competition with others.[11] Dennett rightly points out that the application of Kant's or Mill's theories to an urgent moral problem is unrealistically lengthy and cumbersome. But Dennett's proposal, though an alternative to traditional theoretical ethics in one way, is eminently traditional in another—it too takes quandaries to be the paradigmatic moral situations, and it takes theory, albeit evolutionary, to be the solution.

Can the evolutionary *is* replace the moral *ought*? Dennett gives a well-qualified no and yes for an answer.[12] Most proponents of evolutionary theory, when directly bumping into the question, reply with a clear no, though the pious truism of "you can't derive ought from is" frequently suffers confusion and violation in the particulars of evolutionary arguments.[13]

If we are really creatures of evolution, should not our ability to say "you ought to treat women as the equals of men" have evolved too? Must not the moral *ought* somehow be embedded in the evolutionary *is*? If a human faculty falls outside evolutionary explanation, we seem to be caught between either having to restrict the scope of evolutionary theory in implausible ways or having to deny the elusive faculty, for example the faculty to establish and follow norms of conduct that override whatever may be the case and the case may be. There is, however, a way of passing between the horns of this dilemma.

Consider a peacock's tail feathers. To what sort of environmental circumstance could so extravagant an ornament possibly be adapted? The relevant circumstance is not in the peafowl's environment, it is rather the peahens. They are attracted to the bearer of the largest and most colorful fan. But how could an otherwise so useless and even bothersome attachment attract females in the first place? Presumably the male tail was at one time in evolutionary pea history far more practical, and yet, no matter how short and drab at first, a relatively big tail of vigorous color bespoke health and strength—desirable traits in a mate.[14] From then on it was an arms race among males to top a big and colorful tail with a yet more expansive and shimmering one, and there was a parallel rise of connoisseurship among peahens.[15]

Granted that the peacock's tail and its function fit into the sexual competition part of evolutionary adaptation, it so happened that, regardless of evolutionary considerations, the tails of peacocks became aesthetic delights, symmetrical displays of captivating colors.[16] Thus pea aesthetics transcends evolution without violating it.

Something like this must have happened in the case of human thinking and speaking. The emergence of the most rudimentary ability to consider and convey the meaning of things emerged from random variation and environmental selection to equip humans with a trait that gave them a special fitness and survival edge. At first prominent acuity and expressiveness were both environmentally helpful and sexually attractive. But just as the peacock's tail began to exceed the requirements of balance and navigation and to extend into the aesthetic realm, so human thought and language surpassed the useful and began to expand into the true, the good, and the beautiful.

The rules of arithmetic are not derivable from evolutionary theory. But evidently evolution through sexual selection sharpened human understanding to the point where humans were able to discover and articulate the world of numbers, magnitudes, and their relations. Similarly humans came to grasp the norms of goodness and beauty. These too are not derivable from evolutionary theory, though they have stronger and deeper roots in evolutionary development. Thus humans can transcend, without violating, the realm of evolutionary utility and description and rise to the realm of norms—the norms of rigor and consistency in mathematics, the norms of moral excellence in ethics, and the norms of beauty and artfulness in aesthetics. Evolutionary theory does not determine ethics, but neither does it leave ethics undetermined. Philosophers speak of evolution as underdetermining ethics.

The Mechanics and the Meaning of Evolution

The idea that evolution is underdetermining ethics becomes clear when we distinguish between the mechanics of evolution and the meaning of evolution (or, as the philosophers might put it, between the syntax and the semantics of evolution). The mechanics are the interlocking processes of variation, selection, and continuity and the sequence of life forms they produce. Meaning creeps in without our realizing it. It begins with the question of what the relevant unit of evolution and explanation is. Is it the gene, the organism, a population, or an ecosystem? With each step up in this order, more environmental factors come into play along with social and, in the case of humans, cultural structures that evolve in their own way and constrain biological evolution at their own and at lower levels. What is meaningful and what is mere clutter in this thicket of structures and relations? The answers that are given, most often implicitly, champion one meaning over other possible meanings.

Similarly, when we speak of the survival of the fittest, we think of fitness as excellence and of survival as triumph. But why not think of fitness as adaptation, of adaptation as compromising, and of compromising as giving in? Why not go down in a blaze of glory rather than survive? And do the surviving species deserve to be called "elegant solutions" of the challenges of the environment? Why not say they are "just good enough" to do the job and why not point out that the "higher" organisms are really cobbled together from available structures and organs that were pressed into service as needed? And is complex better than simple? Are humans superior to insects? Copernicus (1473–1543) recommended his cosmological model because it was simpler than the Ptolemaic. He said about his geocentric predecessors: "With them it is as though an artist were to gather the hands, feet, head, and other members for his images from diverse models, each part excellently drawn, but not related to a single body, and since they in no way match each other, the result would be a monster rather than man."[17] Copernicus assumed that the human body and the solar system reflected a grand design. But from a legitimate point of view, the human body just happened. The spine, "excellently drawn" for four-leggeds, was pressed into service as an upright support with monstrous consequences for people with chronic backaches. The reptilian brain, "excellently drawn" for geckos and alligators, on occasion is a monster, at war with the rational cortex in humans.

The mechanism of evolution displays many meanings though there are limits to interpretation. If someone were to see in evolution the lesson of how parents bestow laboriously acquired virtues as inborn traits to their children, the evolutionary mechanism would disallow such a (Lamarckian) view. It also disallows creationism and what has come to be known as "intelligent design." But which of the allowable views should prevail? What *is* the meaning of evolution? It's the view that, when it is well spoken for, will recommend itself. The problem with being an effective advocate of the central meaning of evolution lies in the inevitable weakness of advocacy. As an advocate you plead, appeal, testify, and you hope to reveal. But in the end your listeners must see for themselves a vision that always has alternatives and can be rejected in favor of one of these alternatives.

Not so with the mechanics of evolution. They are what they are, and you cannot reasonably reject them. Hence in explaining them, your discourse is cogent and authoritative. Richard Dawkins revels in the power of this language and delights in savaging dissenters. But he is invariably mistaken in thinking that when it comes to the meaning of evolution, he wields the same compelling force of argument on behalf of his favorite view.[18] In any case, what is the central meaning of evolution? Like the peacock's tail, it is a display of grandeur and beauty, and as for humans, Blaise Pascal (1623–1662) anticipated its work—evolution has placed humans between the brutes and the angels, between misery and grandeur.[19]

What does this mean practically? As Robert Wright has impressively argued, evolutionary theory reveals background conditions of the good life that we ignore to our detriment.[20] Evolution does not sanction male sexual aggressiveness, but it alerts us to a natural tendency we better be aware of. Evolution does not enjoin altruism, but it gives us the assurance that humans are not by nature selfish and belligerent (though clearly self-centeredness and combativeness are things we are also capable of). Evolution does not yield zoning laws and building codes, but it does suggest what kinds of environments humans are likely to prosper in.

In summary, then, the meanings and background conditions that we should take from evolutionary psychology and bear in mind when we are trying to determine the good life are these: First we should realize that the process of evolution has unfolded a world of grandeur and beauty that harbors but is not overcome by violence; second we have to remember that humans have evolved in this process under certain conditions (the ancestral environment) in which they came to prosper materially and morally; and third it must be understood that evolution has given us the talents to discover norms of truth, goodness, and beauty.

The first realization together with the third teaches us that we can be a violent tribe and need to curb that inclination through friendship and justice when it comes to fellow humans and through stewardship as regards the natural environment. The recollection of the second issue should lead us to ask whether we have not moved too far, in the tangible arrangements of our lives, from the circumstances wherein we knew how to do well and feel well. If the answer is yes, the virtues of economy and design are needed to recognize and change these arrangements. In these ways, evolutionary psychology, like Kantian and utilitarian ethics, contributes pieces to the skeleton of a good society but fails to give us the living tissue of the good life.

CHAPTER EIGHT

John Rawls

Principles of Justice

John Rawls (1921–2002) has a prominent position in real American ethics. First of all he is, in my view, the finest American philosopher of the last century, the first to give American philosophy a rank equal to American power and culture. Next, his style is measured and generous, a welcome contrast to mainstream American philosophy with its frequently dense prose and combative tone. Finally, his philosophy is resourceful in its scholarship. Rawls's great book, *A Theory of Justice* (1971), has been an elementary school of moral philosophy for many of us.[1]

Rawls is a person you can admire, and you should. He was tall and athletic, an exceptional baseball player in his youth. He had an engaging stutter and a total lack of affectation. He was solicitous to colleagues famous and obscure, a devoted teacher, a sterling man, a member of the elite. But his crucial trait was this: Unconditional and genuine compassion for people in need. Given his reserve, Rawls's charity was disguised with understatement and learning. Yet it is unmistakable.

As for its substance, Rawls's political ethics is truly American because it captures what's best in the soul of our body politic: the commitment to freedom, to openness, to generosity. Rawls's work also took an important step in the direction of real ethics. Rawls spelled out what may be called the social version of Churchill's principle: We shape our social institutions; afterward our social institutions shape us. The fact that Rawls was entirely oblivious to the original version of the principle, the one that is concerned with tangible reality, is the great limitation of Rawls's philosophy.

Rawls saw his enterprise as a renewal and development of Kantian ethics. Not surprisingly then, Rawls's first concern was to honor and protect the dignity of every individual. Right at the start of the book, Rawls put it this way: "Each person possesses an inviolability founded on justice that even the welfare of society as a whole cannot override."[2]

This concern for the individual's dignity was spelled out in terms of the other two Kantian landmarks—equality and liberty—as the first principle of justice: "Each person is to have an equal right to the most extensive scheme of equal basic liberties compatible with a similar scheme of liberties for others."[3] Rawls followed Kant in conceiving of equality as a moral rather than material issue. Hence equality of material circumstances for each person or household is not required by the first principle of justice. At the same time, inequalities of income, wealth, and consequently of political power and social standing can be terribly injurious to the dignity of the poor and powerless. Rawls's task, therefore, was to find a concept of inequality that was bounded by the moral concern for equality, liberty, and dignity. Put differently, the question Rawls had to answer was this: What kinds of material and social differences among individuals are acceptable from the standpoint of justice? Rawls's reply was the celebrated difference principle, the second principle of justice, the one that embodies the spirit of openness and generosity: "Social and economic inequalities are to be arranged so that they are both (a) to the benefit of the least advantaged and (b) attached to positions and offices open to all under conditions of fair equality of opportunity."[4]

Rawls thought we would see the wisdom of these two principles if we set aside our particular advantages or burdens of natural endowments and social status and imagined what principles of justice we would choose if we considered ourselves exclusively as equal and rational members of an assembly, each concerned about his or her welfare without wanting to pursue one's well-being at the expense of others. This situation of the primal choice of principles of justice Rawls called *the original position*. In such a situation, Rawls reasoned, you would realize that of all imaginable social scenarios the one of greatest concern to you is the situation where you are weak, ill, poor, and plagued by misfortune. You would therefore choose the social design that best protects you in such circumstances—Rawls's principles of justice.

Ballet and Basketball

You can take the life of a society to be a ballet or a basketball game. In the first case, everything is artfully scripted and, if well executed, can make for a wonderful spectacle. In the second case, you lay down certain rules and let her rip. Ingenuity, improvisation, taking advantage of the lanes that open up, all this generates, if all goes well, an exciting game. Rawls realized that basketball is more American than ballet. More important, ballet is analogous to doctrinaire socialism, and it has turned out that the economic life of a society is far too complex to follow a five-year plan that scripts the assignments of labor, the allocation of resources, the goals of production, the distribution of goods, and so forth. A so-called command economy inevitably leads to bottlenecks, shortages, misallocations, and the daily absurdities that the black humor of communist societies used to pillory.

The tribulations of socialism show how formidable the tasks of a well-functioning economy are and how amazingly efficient the market is in solving these problems. The market of course has its own economic liabilities. Left to itself it can result in disastrous oscillations between boom and bust, and it can degenerate into bandit capitalism—the rule of the ruthless—or crony capitalism—the rule of the well connected. Even a carefully hedged market system like ours has fluctuations whose downsides fall disproportionately on the poorly educated, the workers in sunset industries, on single mothers, and on ethnic minorities.

Rawls wanted to honor the energy and ingenuity of an economy patterned after a game, yet he also wanted to draw the economic constraints more fairly and narrowly than the ones that regulate our present economy. These constraints are the social realization of Churchill's principle. Rawls saw that you cannot, no matter what Kant thought, give people two or three principles and tell them: Just go by these principles, and you will be all right. Reality is too complex and people are too weak for this to work. Reality overwhelms principles with complexity, and people subvert them in their moments of weakness and stress.

We need something more robust and reliable, social structures that broadly bank and channel daily life the way the rules of basketball contain and shape the game. Rawls called these social structures the "background institutions of a just society."[5] The first of these is a free and fair political system. Next comes an educational system that gives everyone comparable skills and opportunities. The economy too is to provide free and equal opportunities; and, as a support of the last resort, the government must secure a minimal standard of living for all. These institutions would realize Roosevelt's Second Bill of Rights.[6]

To establish and maintain these institutions, government will have four branches. The allocation branch provides for competitive markets and widely distributed economic power. The stabilization branch is to assure "reasonably full employment." The transfer branch provides the "social minimum," that is, a floor of prosperity for all via a guaranteed minimum income. The distribution branch keeps the differences of individual prosperity within bounds.

Once these institutions are in place, the determination of the good life is left to the individual. Rawls shared with Kant and Mill the conviction that the individual as a rational and moral person is the ultimate ethical authority.[7] To honor individual dignity and autonomy, government must not prejudge what individuals will decide is best for them, and it must therefore restrict its sphere of action to things that anyone will need and want whatever the person's conception of the good life. Rawls called these things primary social goods and says they consist of rights and liberties, opportunities and powers, income and wealth, and what is needed for self-respect.[8]

Justice and the Good Life

Like Mill, Rawls seemed to be worried about whether individuals would in fact make wise choices in charting the course of their lives. As if to reassure himself, Rawls posited the efficacy of what he called "the Aristotelian principle." "The intuitive idea here," he said, "is that human beings take more pleasure in doing something as they become more proficient at it, and of two activities they do equally well, they prefer the one calling on a larger repertoire of more intricate and subtle discriminations. For example, chess is a more complicated and subtle game than checkers, and algebra is more intricate than elementary arithmetic. Thus the principle says that someone who can do both generally prefers playing chess to playing checkers, and that he would rather study algebra than arithmetic."[9]

You probably have to be a Harvard professor, surrounded by ambitious colleagues and bright students, to believe this. Most people in this country are literate enough to read good fiction, but they watch television; and when they do, a serious discussion or a play by Shakespeare would call "on a larger repertoire of more intricate and subtle discriminations" than would situation comedies and "reality" shows, but people overwhelmingly watch the latter. Most people can study the recipes of a fine cookbook, but instead they study the menu at McDonald's.

Something is evidently missing in Rawls's political ethics. His theory is significantly incomplete. Any theory is incomplete, of course, and criticizing a book for what it fails to treat is a cheap way of scoring points. A theory is significantly incomplete, however, if it overlooks something that is importantly within its declared scope. In the last statement of his theory Rawls said that it "takes the primary subject of political justice to be the basic structure of society," and that structure, he said, "is the background social framework within which the activities of associations and individuals take place."[10]

The first thing that is missing in Rawls's theory is the realization that the primary social goods are always culturally biased; they are not "all-purpose means."[11] What comes to mind when we try to think of structures provided by the government that can be used for all kinds of purposes are systems of transportation, commerce, and information. The government presumably should build an efficient highway system, but not tell you where to go. It should favor malls and supermarkets, but not restrict what one may buy. It should secure access to the Internet for anyone, but not influence what will be transmitted. All these institutions are means toward a variety of lives. The question arises, however, about whether these opportunities are unbiased as regards a person's or a community's conception of the good life. A moment's reflection tells us that an elaborate highway system militates against public transportation and walkable cities, that malls and supermarkets have drawn vitality from city centers, and that the promotion of the Internet and easy use of e-mail reduces face-to-face meetings and tangible engagements.

Collective decisions are inevitably biased. This is concealed from mainstream political thought for two reasons. First, some of our common decisions are biased in the right direction; they are rightly acclaimed and hence uncontroversial decisions that pertain, for example, to public health. No one would think of reconsidering or unmaking them. The second reason why the cultural bias of our social decisions has largely remained hidden reveals a second significant oversight in Rawls's theory. The idea of wide individual choice can be given a semblance of realization in the consumer society. Thus the superficial and to my mind detrimental affinity between Rawls's theory and present cultural reality protects contemporary culture from critical scrutiny. One way of piercing these concealments is to follow up on the moral meaning of commodification, the concern of a later chapter.

The final blind spot in Rawls's vision is brought out in the remarks the editor of the last version of Rawls's theory made on the back cover of *Justice as Fairness*: "Rawls is well aware that since the publication of *A Theory of Justice* in 1971, American society has moved farther away from the idea of justice as fairness." We should at least suspect that this growing indifference to social justice has as much to do with the character of the material culture we have been busily putting in place over the last generation as with the individual decisions people have made in the polling booth.

In any event, the point is not to reject Rawls's enterprise but to complete it and to employ the original version of Churchill's principle in addition to the great social and political version Rawls has provided.

PART TWO

Practical Ethics

CHAPTER NINE

Theory and Practice

Applying Theory to Practice

Theory is king when it comes to ruling reality. We think of theory both as the skeleton that gives a thing its shape and makes it move the way it does, and as the grasp of that structure, the understanding that enables us to take control of a thing. Take illness. A nurse may be good at wiping the brow of a cancer patient, but it will take a breakthrough in medical research, a better theory of what underlies cancer, to heal a patient with a melanoma or lymphoma. That's the conventional justification for why nurses make five-figure salaries and research professors in medical schools seven-figure salaries.

A theory is a set of laws, principles, or rules plus some stage setting, explanations of what the laws or regularities mean and of how they bear on reality. The laws of physics are the epitome of what we expect at the center of a theory, laws of universal scope, rigorous precision, and illuminating force. The outstanding intellectual challenge of today is to discover the "unified theory," a mathematical theory that would unite the inconsistent pillars of contemporary physics—relativity theory and quantum theory—and specifically explain the origin, structure, and fate of our universe.

Practically, such a theory would be useless. It would be a sheer intellectual triumph. But many more limited physical and chemical theories have had enormously useful applications. Most everything we use and consume today owes its helpful or commodious character to scientific theories. The theoretical substructure of contemporary culture does not, of course, consist of clean and elegant scientific theories. It is more like a framework of trusses, with members extending and connecting in lots of ways. Still, even in mundane matters, we expect our problems to be solved on the basis of principled insight that connects up to basic scientific laws. Say your car suddenly breaks down for no apparent reason. You take it to a mechanic, and when you pick it up, you ask: "Did you find out what was wrong with it?" The mechanic answers: "I tried different things. All of a sudden it started; it's running fine now. But don't ask me why." You would not consider this a satisfactory job. We expect our cars to be repaired by the principles of auto mechanics, and we expect those principles to be consistent with all relevant principles of physics and chemistry.

What goes for physics should hold for ethics too. That's the general expectation. Consequently there has been in the last quarter century a call to apply ethical theory to practical problems and circumstances. Philosophers have been asked to advise doctors, lawyers, journalists, and other professionals on their codes of conduct and on moral quandaries. This is by and large a development we should welcome and support. Awareness of ethical issues and recourse to reasoned discussion and agreement are important features of a decent and virtuous society.

Just as important, solving problems through principled discourse is eminently in keeping with the character of the institutions that chiefly have invited such discourse—the professions. The hallmarks of a profession are self-regulation; responsibility for a precious social good such as health, justice, or education; and, most relevant for our purposes, devotion to public principles. The rise of professional ethics is a late and welcome fruit of the devotion to the Enlightenment norms of equality, dignity, and self-determination. That sense of moral responsibility becomes especially important when there is pressure to push professional issues out of the jurisdiction of the professions and into the market.[1]

Professional concerns shade over into social ones, and thus professional ethics is contiguous with practical ethics. The application of ethical theory is particularly effective and impressive where problems are well articulated and we can be confident that answers will be forthcoming. Definite problems, so-called quandaries, are in fact the native soil of theoretical ethics. Kant introduced and illustrated the moral law through common quandaries.[2] And Mill, having regaled us with quandaries concerning punishment, just wages, and fair taxation concludes: "From these confusions there is no other mode of extrication than the utilitarian."[3]

These days information technology generates conundrums that fall within established moral theory and will likely be resolved by it. How do we define and enforce copyright in cyberspace?[4] Does my employer have the right to examine my computer use at will? Is a software firm liable for the damages that follow from bugs in a program it has written and sold?[5] Not that the answers are always easy and obvious, but we trust that in light of our principles of property, autonomy, and responsibility these questions constitute solvable puzzles.

In other regions of practical affairs, the questions are clear and urgent, yet we are anything but confident of answers. In earlier times, nature answered these questions so firmly and regularly that we were hardly aware of the questions and did not think of questioning the answers. Half a century ago, a baby born prematurely after twenty-six weeks of pregnancy could not live for long, and we had to accept her death. Today she can be kept alive and nursed to self-sustaining life, but at some risk to normal development. As medical technology progresses, ever less developed babies can be saved and grow up to ever more questionable health of mind and body.[6] Or should we draw the line and save no baby under thirty-two weeks of pregnancy? And if so, who is to hold the line, the doctors? Possibly against the fervent wishes of the parents?

Progress in science and technology has given us new powers in this instance, but, it seems, no guidance for how to use them. We have gained similar powers and suffer similar quandaries when it comes to the end of life. And then, before long, genomics and genetics promise to give us dominion over the genetic quality of a person's entire life. Forbidding complexity of genetics and the environment will most likely keep a person's character from our manipulations. But simple features of health and physical appearance may well be ours to shape. How are we to wield such power?[7]

One answer is to refuse dominion and to leave decisions to contingency or providence. Of the Siamese twins, Jodie (or Gracie) and Mary, born August 8, 2000, in Manchester, UK, Jodie could live a normal life if separated from Mary who, dependent on Jodie's heart, would die. If not separated, both would soon die due to the insufficiency of one heart for two humans. The judiciary urged separation. The twins' parents demurred, not wanting to interfere with God's will.[8] But, to put things religiously, did not divine providence give us the power and responsibility to decide? Is not the refusal to decide a way of deciding too? In the event, the court prevailed. Mary died, as anticipated, and Jodie by now is a healthy girl.

If our concern is the life of virtue and excellence, then, hard as they are, these questions are both too sharply articulated and too preliminary to reveal the nature of the good life. Assume that all the quandaries mentioned so far had been solved satisfactorily. We would then barely stand at the threshold of the good life. Nothing so far would have been said about what it is to lead a vigorous, accomplished, and blessed life. This is true of pretty much all the topics that come under the rubric of practical ethics today, the social, political, and environmental issues that unfold into the problems of racial and sexual discrimination, euthanasia, the death penalty, abortion, animal rights, wilderness protection, world hunger, and all the other problems you find in textbooks on practical ethics.[9]

The efforts to solve these problems deserve admiration and support, and, having solved these problems, we would have a better chance of leading the good life. But getting stuck in possibilities is the curse of contemporary ethics, and succumbing to banal actualities is the curse of contemporary life. While it is true that the solutions of practical ethics enhance the prospects of a good life, the reverse is true too. Knowing and living a life of moral excellence casts light on how life should begin, should it be shared, and should end.

Moral Practices

What the tissue of skin and flesh is to the human bone structure, practice is to ethical theory. Theories are hard and austere; practices are soft and rich. Theories are clear and precise; practices blend with one another and are ambiguous. Yet like the flesh that clothes a human skeleton and like its most expressive region, the face, practices have identities and character. They are ethically significant and sometimes striking. But again, like the human face, the significance of a practice cannot be reduced to a set of formal features. Consider how hopelessly overwhelming a task it would be to explicate and formalize the texture of skills and circumstances that make up a particular practice.

What might the practices in the daily life of the protagonist in Kant's story have been like?[10] At the end of the eighteenth century in East Prussia, he may well have been a day laborer who went to the village square early in the morning hoping for work on that particular day much as migrant workers in this century do today. Assume the foreman of an estate came by, looked at his face and his health and hired him to help dig a ditch. Our man agreed to a daily wage and followed the foreman to the digging site, received a pick, a shovel, and instructions, and began to work.

Consider all the skills the laborer had to draw on and, most important, how doing the right thing is indivisibly a technical and ethical enterprise. He had to judge whether the foreman had in fact work to be done and could be trusted to pay him at the end of the day. He had to appraise the fairness of the wage offered, given the alternatives and the kind of work he was about to accept. When digging, he had to swing the pick in short and rapid strokes where the soil was soft and in sweeping, over-the-head swings where it was hard. Using the shovel, he would, if he was right-handed, put his left foot slightly forward and the left hand lower on the handle. He would terminate the arc of the shovel over his left shoulder so the dirt would sail as a compact clump to the intended spot.

Swinging the pick indifferently or letting the dirt slide off the shovel in a diffuse spray would be the wrong thing to do, but wrong in what sense? It could be so in the sense of ignorance, laziness, distraction, or spitefulness, and the foreman in turn had to be able to ascertain the cause of the poor performance and to instruct, admonish, or dismiss the man. How grandiose and unhelpful would it be to proclaim: The day laborer entered into an obligation, and he was duty-bound to fulfill it. So many crucial conditions and qualifications are covered over. Can you enter an obligation if your skills are deficient? Is it enough to make a good faith effort? Or is it reprehensible to be ignorant of one's lack of skill or endurance? More important, how wrong-headed and futile would it be to try to envision and regulate all possible circumstances in advance. What if the handle of the shovel was only two feet long? Could the laborer walk away from his obligations? What if he had a severe cold? Would he have been disqualified? What if a heavy rain started to fall? All such circumstances inform our moral situations and obligations. It is impossible to foresee them all, and it is less possible to anticipate the ways they qualify each other.

Practice has been used to remind us of this complexity and to serve as a guide to the patterns of our conduct. But such work is always in danger of using *practice* or similar terms as an argument stopper or court of last appeal. In that vein, to show that something is a practice is supposed to clinch an argument. Recourse to practice, tacit knowledge, weltanschauung, paradigm, ideology, presupposition, and the like has been criticized on the grounds that practice and its kin are not the definite objects they are made out to be, do not have the causal powers that have been attributed to them, and fail to have the continuity from one generation to the next that is crucial to their identity.[11]

These criticisms, however, amount to little more than showing that a practice is not a theory. Alternatives to theoretical ethics inevitably appear to be poor substitutes when one remembers the definite, lawlike, and compelling character of moral principles. But these traits are austere by themselves and become fruitful only when they get embedded in richer and thicker moral textures. How then can we talk intelligently about practices in their own right? We can do so the same way we talk about types of persons and kinds of faces—the athlete, the scholar, the model, the businessperson, the military type, the outdoors person, the flower child, and such. These categories are both fuzzy and helpful, and they are helpful even when someone does not look the type, as when a scholar looks like an athlete or a banker like a flower child.

Moral practices, again like human faces, show admirable types and types that are pinched, angry, or arrogant. The patterns of excellence in ethical conduct have traditionally been called virtues and their opposites vices, and there was a time when teaching the virtues played a prominent role in the higher education of this country. In the nineteenth century, college students concluded their course of studies with a lecture on ethics, usually delivered by the president.[12] A late and elaborate example of this is recorded in *Practical Ethics* by William DeWitt Hyde, who was president of Bowdoin College from 1885 till his death in 1917.[13] He completed the writing of his book in 1892 when he was only 34.

President Hyde saw the problem of moral instruction much as it appears today. Between theory, "excellent mental gymnastic for the mature," and practical advice, "admirable emotional pabulum for the childish mind," as he put it, he saw the need for a book that "must have theory; yet the theory must not be made obtrusive nor stated too abstractly."[14] Hyde could not foresee that his kind of moral instruction was soon to be extinguished entirely by theory. At the turn from the nineteenth to the twentieth century, philosophy, like law and medicine, became professionalized and was taken out of the hands of dilettantes like college presidents.[15] Hyde, for example, had "only" a doctor of divinity.

Hyde's proposal was not just the victim of professorial arrogance. It suffered from substantial flaws. Entirely in the modern spirit, it aimed to be systematic and comprehensive. It was outlined as a matrix in which the first column consisted of twenty-three objects, ranging from food and drink to God. With each object a duty was associated; with each duty a virtue; with each virtue a reward, a temptation, and, in the spirit of Aristotle, a vice of defect and a vice of excess. The last column contained the penalty that you had to expect should vice prevail over virtue. Thus President Hyde tried to capture and tame the moral life with a grid of twenty-three by eight, that is, 184 items.

Hyde's attempt to fill the gap between theory and practice remained too much in the shadow of theory to answer the complexity and confusion of the coming century. But the void he saw between theory and practical advice has remained open to this day. For we have today not only an abundance of treatises "which are exhaustive in their presentation of ethical theory," we also have much "pabulum for the childish mind."

Not all nonphilosophical moral instruction is pabulum, not, for example, Harold Kushner's admirable *When Bad Things Happen to Good People,* nor the sensible weekly columns Randy Cohen writes for the *New York Times Magazine*.[16] Every week the best-seller lists contain one or two books that, with an unearned sense of superiority, oblige the common hunger for moral instruction with the claim that the successful life is really very simple, that it has eluded people only because they have overlooked either one big thing—the inner child or the 75 percent of their brain that is not used—or they are ignorant of a definitive list of seven or thirteen habits or attitudes.

Contemporary virtue ethics promises moral guidance that takes a more comprehensive view of morality than theoretical ethics, has more scholarly substance than Hyde's *Practical Ethics,* and greater moral seriousness than the offerings of gurus. Virtue theorists have rightly pointed out that their approach is not only richer and more encompassing than ethical theory, but that it also goes beyond the austerity of moral obligation and considers moral excellence; and finally they stress that virtue ethics is not codifiable as a set of laws or principles. Most of their intellectual energy has been spent, however, not on moral instruction but on distinguishing virtue ethics from the ethical theories of Kant and Mill, that is, deontology and utilitarian teleology.[17]

Equality, Goodness, and Excellence

We have mixed feelings about the pursuit of excellence, feelings of the heart and feelings we express in public. In our hearts we feel the claims of excellence, the challenge to be and to do the best we can; we feel disgust and sorrow when we chicken out. Our failures also make us resentful and make us think up reasons why we should be free to do as we please and remain independent of the claims of excellence. Still, we can hardly deny that in American society, certain sorts of athletic, artistic, and managerial excellence are well recognized and rewarded with ungodly sums of money. Below these stars, there is a lesser world of meritocracy where the amount of fame and fortune you get is determined by the degree of ambition, intelligence, and good looks you possess. Our educational system too reflects an aristocracy. It extends from the upper reaches where you find Harvard to the lower reaches where you find, should you be looking for it, Miles City Community College.

We are reluctant, however, to acknowledge the hierarchy of talents. Conservatives are content with their silent belief that they are the rightful beneficiaries of their gifts. Liberals would find it politically incorrect to point out that some people are less intelligent and diligent than others. It was not always so. In January 1943, President James Bryant Conant of Harvard University appointed a faculty committee on "The Objectives of a General Education in a Free Society." The result came out in 1945 and is known as the Red Book.[18] It contained thirteen pages on "Kinds of Difference." Those were divided into "an inner sphere of ability and outlook and an outer sphere of opportunity."[19] Although the authors stressed the interplay of the two spheres, they nonetheless concluded: "The best schools and most modern housing do not suddenly endue all the young people within them with high standards and good ability."[20] As for the prospects of the less talented, they had this to recommend: "But young people of average intelligence, though not suited for the traditional college, can yet profit from training in agriculture or nursing and from many kinds of courses, largely vocational, offered by junior colleges and technical institutes. It is evidently as important for their welfare and that of society that they make the best of themselves as that the more gifted do so."[21]

Literature is not mentioned by the authors of the Red Book as something "young people of average intelligence" may "profit from." That is surely a prejudice of intellectuals.[22] More broadly, a substantial general education program, though admittedly not advanced work in mathematics, physics, or philosophy, is within everyone's grasp. But such questions were in any case buried when talk about the "outer sphere of opportunity" swamped consideration of the "inner sphere of ability and outlook."[23] Rightly so, one may add, because equality and excellence or democracy and aristocracy are incompatible as social and political goals. Equality and democracy have priority since from the moral point of view we are all equals. Philosophers have elaborated this point by distinguishing between blamelessness and excellence, and, if excellence you must have, between moral and nonmoral excellence.[24]

Moral excellence is governed by the will and pertains to those things that are subject to your will—honesty, generosity, courage, diligence. If you are lazy or a liar, it's your fault. Not so with the nonmoral qualities such as intelligence, good looks, musical talent, or athletic ability. It's not your fault if you are tone-deaf or a stumblebum. Thus moral excellence is compatible with equality. You too could be morally excellent if only you wanted to. That's the conventional view.

At first sight, the two kinds of excellence or goodness seem to cleave apart neatly. Consider the Sinatra problem. Frank Sinatra was an excellent singer, one of the great stylists of the last century. But was he a good person? Hans van Meegeren was a good forger of Vermeer paintings, but he was found guilty of forgery and fraud and sentenced to a year in prison. The ways we use the word *good* reflect a distinction between moral goodness and nonmoral goodness, between the goodness of the Good Samaritan and the goodness of the good forger. There is also the case, so it seems, of the person who is morally good, but not good at what he does, say a good-hearted carpenter who does lousy work.

This distinction leaves the real problem unsolved. According to the liberal view, John Rawls's for example, you deserve credit for your moral excellence, for your honesty, diligence, and generosity because they are subject to your will, but not for your nonmoral, that is, natural and cultural, excellence, for your musical talents, your mathematical acumen, or your leg speed because you did not and could not will to have them. But is willpower any less of a natural gift than intelligence? And yet, officially, conservatives no less than liberals applaud candor and industriousness and would not think of deploring lack of intelligence or beauty. Why do we cling to the moral-nonmoral distinction?

The distinction sanctions social injustice and protects us from the uncomfortable task of examining and reforming our questionable notions of the pursuit of excellence and happiness. As for social justice, the rich and powerful do not invariably owe their privileges to the possession and exercise of natural and cultural excellence. Some were plain lucky. Their business took off because conditions happened to be explosive. Or they are the sons or daughters of the rich and well connected. If we leave the grounds of privilege dark, few will complain.

To the extent that we recognize excellence and reward it in a way we think is really sincere, that is, with money, it is a cramped and crabbed kind of natural and cultural excellence. And what is perhaps worse, the rewards of excellence that we so sincerely and earnestly bestow or envy, are not truly rewarding at all. Our refusal to examine and reform and then promote other norms of natural and cultural excellence also disburdens us of the challenge to consider how Churchill's principle bears on the quality of ordinary life. As a consequence life is so often more sullen, disengaged, and unfulfilled than it needs to be. But there remains this deep and honorable misgiving we have about the acknowledgment and promotion of nonmoral excellence. Celebrating natural and cultural excellence seems to expose so many of us as duds and losers. It seems unjust, a violation of what we have fought hard to establish—equality.

Thus we are stuck with the task of seeking a common root of moral and natural or cultural excellence. We in this country, no matter what our European critics say, have a fair claim, if not to moral excellence, at least to moral decency. When it comes to standards of cultural excellence, however, we surely fail to deserve a grade even of fair. We are not nearly as knowledgeable, as athletic, and as artistic as our opportunities permit us to be. Yet if the foundations and standards of cultural excellence are, as if by definition, nonmoral, then we have no obligation to pursue cultural excellence and no right to expect of one another that we dedicate ourselves to it. Moral and nonmoral goodness obviously can and often do cleave apart. But can they become one, and, in the ideal case, are they in fact one?

There are two ways of beginning to bridge the gap between moral excellence and natural and cultural excellence. One way of acknowledging superior talent and of encouraging its development while upholding the claims of equality is, in Rawls's words, to "view the greater abilities as a social asset to be used for the common advantage."[25] A religious reply to the same effect is to regard our talents not as tickets to success and acclaim but as obligations to benefit our sisters and brothers. Another way of connecting moral with natural and cultural excellence is to remind ourselves that the latter rarely exists without the former. Sheer talent is not usually praised and rewarded. Talent needs to be developed if it is to grow into excellence, and such development requires the moral excellence of hard work and self-discipline.

Creditable as these answers to the questions of excellence are, they nevertheless are second best. They make room for excellence alongside moral goodness, but they fail to uncover a common root or reveal a common fruit. Do moral goodness and worldly excellence hang together? Let's look at the good-hearted but inept carpenter. Is he a truly good person? It cannot have escaped him that his work is no good. Why is it bad? Is he careless, distracted, or just poorly trained? The ethics of virtue would require him to find out and to remedy the deficiency. If he fails to be concerned, he is not only a bad carpenter but also a bad person. But what if he is devoted to his work, has tried to learn from the best masters, and still cannot do work that customers applaud and appreciate? Ethics requires us to live within the bounds of our talents. To indulge overweening ambitions at the expense of other people is unethical. Thus moral goodness and worldly excellence are ultimately inseparable from each other. There is no consistency of moral excellence with nonmoral bungling.

But there seem to be obvious conflicts of nonmoral excellence with moral excellence. Consider the Sinatra problem. Frank Sinatra was not a good person, yet few connoisseurs of jazz and popular music would deny that Sinatra at his height was one of the great singers of the last century. Is there a resolution to this inconsistency of nonmoral excellence with moral nonexcellence? Sinatra was devoted to his craft—a morally commendable attitude, and perhaps we can say that when he sang, when he engaged his listeners and made them understand life's sorrows and pleasures more deeply, that those occasions were moments of redemption. They were, at any rate, when he sang something like the rueful "In the Wee Small Hours of the Morning." They were less so, if there was redemption at all, when he sang the egomaniacal "My Way." The redemptive unity of moral and nonmoral excellence is open to writers as well. Pico Iyer has said of Somerset Maugham that "this often taciturn and muffled man with the down-turned mouth somehow did what many writers long to do, reserving his meanness and cattiness for life, while getting the best of himself into his art."[26]

This leaves us with the conflict of nonmoral excellence and moral equality, exemplified by the Battle problem. Kathleen Battle was a diva and behaved like one. She never let us forget that she was a star and we were not. And she was. She possessed a supernaturally sweet and lovely soprano. But when in a concert hall she gave an audience her most devoted and accomplished singing, the pleasures were strong and shared on all sides. At least on the best occasions, envy and low self-esteem among the listeners and pride and scorn on the part of the star fell away and made room for common celebration. As redemption is the solution to the Sinatra problem, so reconciliation is to the Battle problem. In fact, excellence, unless spoiled by competition or commerce, is a "nonrivalrous resource" as economists would put it.[27] If your excellence at skiing increases, mine does not have to decrease. The pursuit of excellence can be a win-win or positive sum game. When I saw Ingemar Stenmark carve his turns, I was determined no longer to slide mine.

Although there are common roots of moral goodness and extraordinary excellence of skills, and though there are shared fruits of such excellence in celebration, the roots do not pervade the entire ground of our lives, and the fruits cannot fill up life wall to wall. Hence we need to elaborate credible forms of excellence open to all.

CHAPTER TEN

Personal Virtues

Wisdom

Through William Bennett's best-selling *Book of Virtues* and André Compte-Sponville's *A Small Treatise on the Great Virtues,* best selling in France and successful here in the United States, virtue ethics appears to have enjoyed some popular success.[1] But I suspect these books lend themselves better to giving than to reading. We receive the virtues with dutiful respect, but we have a hard time giving them a place in our lives. No virtue has suffered that fate more obviously than wisdom. For Plato and Aristotle it was the highest of virtues, and in the Hebrew Scriptures it has been lovingly praised. But it now is so foreign to our world that it didn't even make it into the books of Bennett and Comte-Sponville.[2]

Consider how much closer to the good life wisdom would take us compared with the guidance provided by moral landmarks. The norm of dignity, for example, is most helpful when we are really straying from it as we are in treating gays and lesbians. It is helpful also when we are honestly confused as we have been in our dealings with addicts, losers, and criminals. But in so many daily decisions about what to give our children, what to demand of our employees, or what to say to a difficult boss, the norm of dignity is just too distant and lofty. A vigorous tradition and practice of wisdom would tell us what our options are and what fork in the road to take. If the insight would not come to me, lacking in wisdom as I am, there would be a recognized sage or wise woman I could turn to.

The practice of wisdom, however, needs a basis to rest on. For

Plato and Aristotle there were realms of order and permanence that could serve as that foundation. More particularly for Plato, wisdom was the greatest of the four cardinal virtues and the skill that distinguished the ruling elite of the ideal state, the philosopher-kings. Below them were the warriors who naturally needed courage, and below them the workers who regrettably, Plato thought, needed temperance or self-control. The virtue that governed the proper relations among the three classes was justice, the fourth of the cardinal virtues. Wisdom enabled the rulers to see and grasp the eternal norms of what is true, good, and beautiful, the ideas, and thus to direct the state in the best possible way.[3]

Aristotle recognized that different people find happiness in different ways, some in pleasure, others in a life of honor and action, and still others in contemplating the order of reality and its divine source of movement and meaning. Wisdom was the skill of being equal to the demands of the contemplative life, and in wisdom, as Aristotle understood it, the highest human faculty, the noblest object of inquiry, and the best kind of life, all became one.[4]

At just about the time when Plato and Aristotle were teaching and writing, one of the sages and editors of the Hebrew Scriptures assembled the wisdom book known as Proverbs. He may also have been the author of poems on wisdom as a virtue and as the child and companion of God. The intimacy with the divine that is wisdom and that Aristotle alludes to is memorably presented in Proverbs:

> The Lord created me at the beginning of his work,
> the first of his acts long ago.
> Ages ago I was set up, at the first, before the beginning of the earth.[5]

The account of wisdom's origin ends with some of the loveliest lines in the Bible:

> when he marked out the foundations of the earth,
> then I was beside him, like a little child;
> and I was daily his delight, rejoicing before him always
> rejoicing in his inhabited world and delighting in the human race.[6]

A millennium and a half later, in the thirteenth century CE, Thomas of Aquino took up the kinship of wisdom with God, and he combined with it the Platonic notion that wisdom is good judgment in the most significant matters we are faced with: "Wisdom means a certain rectitude of judgment in accordance with divine reasons. Now rectitude of judgment happens to be twofold: first, in accordance with perfect use of reason, second, because of a certain connaturality with the issue to be judged."[7]

As an intellectual virtue, wisdom requires devotion to the life of the intellect. In American culture, however, intellectual pursuits and accomplishments have never ranked high. Distinction in our society tends to come from lots of money while in France it seems to follow lots of education.[8] Although philosophers in the United States are not particularly charmed by this state of affairs, they should admit that it is not all bad and better, perhaps, than the adoration of intellectual virtuosity when little of substance in the real world turns on it.

There is, moreover, what Thomas calls the wisdom of connaturality, the insight that comes from the kinship with the nature of things, earned through practical engagements. This kind of untutored and unpretentious wisdom is compatible with American pragmatism, the sort of sagacity possessed by elders of experience and fortitude. No doubt there is such wisdom, but it is now restricted to the narrow scope of personal relations. A venerable aunt or grandfather may have good advice on how to treat a difficult child or despondent spouse. But when it comes to the puzzles of politics or the trials of technology, a long life bravely lived is not a sufficient source of insight.

The real need is not for wisdom as such but for the order and coherence that is its proper object and setting. Wisdom has faded into platitudes or shrunk to unpaid counseling because its traditional concerns and contexts have evaporated. We no longer believe that there are immutable Platonic standards of goodness and justice. Astrophysics has robbed the Aristotelian worldview of substance. At least in public circumstances we no longer have the nerve to invoke Thomas's divine reasons, and we smile at the thought of wisdom playing like a child at the feet of God—we cannot get ourselves to take this seriously.

Early in 1934, T. S. Eliot composed a church pageant and titled it "The Rock." In its introductory meditation the Chorus reminds the audience of the earthly and heavenly order that once prevailed.

> The Eagle soars in the summit of Heaven,
> The Hunter with his dogs pursues his circuit.
> O perpetual revolution of configured stars,
> O perpetual recurrence of determined seasons,
> O world of spring and autumn, birth and dying!

But now "[e]ndless invention, endless experiment" have upset and dissolved that order, and the Chorus in despair asks:

> Where is the wisdom we have lost in knowledge?
> Where is the knowledge we have lost in information?[9]

The information age has driven our losses one step further into data and at the same time has held on to the pious hope that we can find our way back to wisdom simply by integrating data into knowledge and refining knowledge into wisdom.[10]

Although we lack a common cosmic order of moral significance, we do not live in a world that lacks significant structure. To know that structure is less than wisdom; it is more like educated citizenship. Yet without that knowledge, there is no hope of wisdom. To start with the widest scope, astrophysics gives us both a cosmic view and an account of what the world is like at its largest and smallest. Evolutionary theory tells us how life developed on this particular planet and what sort of evolutionary background conditions govern the human condition. A global survey shows how humans have appropriated their world. A survey of United States history reveals the achievements and burdens of the society we live in. Knowledge of democracy, human rights, and the high-minded moral vision of which they are parts gives us an explicit and intelligent grasp of the ethical norms that have a claim on everyone in this country.

It is worth remembering that this core knowledge used to be two-thirds of what in higher education is called general or liberal education. Arts and literature make up the remaining third. It is a distinctive and in fact glorious part of American colleges and universities and reflects our resolve to take joint responsibility not just for the training but also for the education of our students.[11] Increasingly, however, we have been slipping out from under the burden of that resolve.

Harvard is a case in point. In 1943, President James Bryan Conant charged a faculty committee to design "a general education—a liberal education—not for the relatively few, but for the multitude."[12] As the Conant committee found, "the relatively few" at Harvard too needed a better general education program, one that was more focused and civic. The celebrated Red Book of 1945 was admirably principled and generous and left its imprint on a generation of American higher education. Inevitably, revisions became necessary to do justice to women, to minorities, to the environment, to globalization.

These revisions could have proceeded while holding to the conviction that we inhabit a common country and a common world and that an educated American needs to know the crucial features of both. Translating that conviction into a revised curriculum requires selfless work and high-minded enthusiasm. The general decline of civic engagement and responsibility, however, has overtaken the academy too, Harvard not excepted. In 1997 a sharply divided faculty voted to permit some of the normal departmental courses to be counted as general education courses, and as of spring 2003, more than a quarter of the core general education program consisted of such offerings.[13]

A less obvious but more insidious erosion of general education and knowledge has been the shift from substance and content to skill and procedure. Instead of studying American history, you learn "how to think like a historian," and instead of learning the fundamentals of evolutionary theory, you get to meet "the scientific method." Harvard's Core Program now stresses that the Core "does not define intellectual breadth as the mastery of a set of Great Books, or the digestion of a specific quantum of information, or the surveying of current knowledge in certain fields. Rather the program seeks to introduce students to the major *approaches to knowledge* [their emphasis] in areas that the faculty considers indispensable to undergraduate education."[14]

Does the same broom serve to sweep the many different mansions of history or science or literature? Do we really take the same approach to "Fairy Tales and Children's Literature" as we take to "major 18th-century autobiographical, fictional, and philosophical texts that explore the paradoxes of the modern self," to take two examples from Harvard's Core? And what does the approach to "physical processes that formed the Earth" have in common with the methods employed to study "the evolution, over the past three centuries, of our concept of time" (also from Harvard)?[15] In fact what do the approaches of a poststructuralist theorist and those of a new historicist to Shakespeare's plays share if not their content?

It does not seem likely, then, that the resort to procedure and approach lends much commonality or generality to general education. But neither does it seem reasonable to assume that the widespread turn to method is simply due to a mistake on the part of recent reformers of general education. Procedure is in fact an important element in democracy, in the economy, and in athletics. Who gets to be president of the United States, for example, is not determined by divine sanction, heredity, or virtue, but by a procedure, and the integrity of democracy depends in part on the refusal to violate (but also on the willingness to reform) procedure when substance is compromised. The rise of procedure and approach reflects the seductiveness of all-purpose instruments that promise to put us in control of whatever regions and circumstances, and it goes along with the slackening of cooperative responsibility. But there is a glimmer of hope, at least at Harvard. The College, prodded by President Lawrence Summers, has proposed changes to general education that would put more emphasis on science, on global awareness, and on knowledge rather than skills.[16] But no one will claim that the "Report on Harvard College Curricular Review" of April 2004 equals the Red Book of 1945 in vision, cohesion, and courage.

What then is the general level of education in this country? What do people typically know of science, geography, history, and politics? They know more than alarming snapshots occasionally suggest. But on balance, the levels of knowledge are clearly below the needs of an engaged citizenry and below the least foundations of wisdom.

Most disturbing, perhaps, is the fact that the ground beneath the foundations fails to be generally firm. To acquire knowledge today requires sound reading skills and a command of elementary mathematics. About a third of high school students "don't understand elementary algebra, have little conception of probability and can't make simple measurements of the kind required of a beginning carpenter," and about a quarter of all high school students lack the reading skills that are needed for "comprehending a relatively simple passage from a book."[17]

As for science, more than 70 percent of the population knows that light travels faster than sound and the earth goes around the sun. But should not everyone know this? Only half of the population knows that electrons are smaller than atoms and that it takes the earth a year to go around the sun. Shouldn't everyone know this too? Questions about genetically modified organisms, genetic engineering of humans, cloning, and stem cells are inflaming and dividing the public, but only 45 percent of the population can give an acceptable definition of DNA, and only 22 percent are able to define what a molecule is.[18]

In 1988, nearly all Americans could locate Texas, California, and Canada on a map. But less than half could do so for England or Japan, and less than a third could locate Vietnam, Egypt, or Sweden.[19] Almost all people knew (in 1989) what happened in 1776 and who the first president was. Less than half knew (in 1975) who Karl Marx was, and less than a quarter knew (in 1992) that Bosnia and Serbia once were part of Yugoslavia.[20]

Knowledge of politics is most important for citizenship and the common welfare. It is also, if deep, the most inclusive kind of knowledge. Although the knowledge of science, of geography, and of history has each its own dignity, each also has a place and significance within political knowledge and the body politic. Not surprisingly, the picture of common political knowledge is much the same as of the other regions of knowledge. If you make a comprehensive list of questions that pollsters have asked over the past sixty years about different aspects of politics, you find that few people have been able to answer all of them correctly, few were unable to answer any, and the bulge of people are located in the middle between expertise and massive ignorance.[21] But this implies that fewer than half have been able to tell, to use more recent examples, what the Bill of Rights is (1986), who both of their senators were (1985), how large the population of this country is (1988), and that Kuwait is not a democracy (1991).[22]

If common knowledge were sound, wisdom could build on it—not an easy next step since knowledge would have to be overarched by what we, as a people, do not have—a confident grasp of an encompassing order.

Courage

If wisdom has fallen on hard times because the order of reality and the life of the spirit have dissolved, courage has declined because the hardness of reality and the life of daring have softened. Like wisdom, courage needs a certain context and specific challenges to prosper. Remarkably, the great philosopher of courage, Aristotle, already struggled with a shift of the context for courage.

Courage was the defining virtue of the Homeric hero, and Aristotle (384–322 BCE) evidently loved heroic courage above all. When in the *Nicomachean Ethics* he says of the virtuous man that "he would prefer an hour of rapture to a long period of mild enjoyment, a year of beautiful life to many years of ordinary existence, one great and glorious exploit to many small successes," he surely had in mind Achilles who lived and died by this rule.[23] When it comes to defining the hero's courage, Aristotle ranks courage according to the dangers the hero meets, mortal danger being the severest test. He goes on to ask: "What form of death then is a test of courage?" He answers: "Presumably that which is the most beautiful. Now the most beautiful form of death is death in battle, for it is encountered in the midst of the greatest and most beautiful of dangers."[24]

Aristotle must have been thinking of the exuberance and athleticism of heroic, warrior-to-warrior combat.[25] Translators invariably play down Aristotle's aesthetic delight in war by staying away from the "beautiful" meaning of the Greek word *kalos* and translating it as "noble," "fine," "admirable," and the like.[26] In any case, warriors in Aristotle's time no longer fought for glory and spoils but to protect the city, the women, and the children. And there was little beauty, if some nobility, in the famously effective Greek phalanx. The main concern was to keep those heavy shields locked to make an impenetrable wall and to advance steadily and, if possible, unstoppably.[27] Although Aristotle makes no explicit distinction between heroic and civic courage, his sympathies are evident, and he wistfully concludes that civic courage belongs among the lesser kinds of courage although it is first among them "since it most closely resembles true courage."[28]

Like Aristotle, Thomas of Aquino (1224–1274) was able to look back to a heroic age when courage was a defining and unequivocally glorious Christian virtue. Persecution in the Roman Empire was the ultimate test of faith and hope, and to suffer martyrdom was to pass the test triumphantly. But Thomas was just about as far from the time of martyrs as Aristotle was from the time of heroes—about a thousand years. By the thirteenth century, Christianity was secure and flourishing in Europe. To lead a faithful life no longer required a confrontation with torture and death, but rather the daily struggle with the hardships and distractions of everyday life.

To agree with Aristotle, whom he simply called "the Philosopher," was nearly as important to Thomas as being faithful to Christian doctrine. Hence he did not want to deny that war is the preeminent setting of courage, but he widened the meaning of war to include particular attacks on one's life or well-being. Moreover, he divided courage into an aggressive and an enduring component, and appealing to a throwaway line of Aristotle's, he elevated endurance over acts of daring.[29]

As William Miller tells us in a superb account of courage, David Hume and Adam Smith, who saw the modern refinement of manners and the promise of technological comfort, began to worry about the fate of courage.[30] The dangers and hardships that made courage prosper were receding, but the need for the moral vigor based on courage continued. For more than three hundred years now, modern technology has been dissolving traditional hardships and indulging our weaknesses. But every generation, or at least every century, rediscovers the need for challenges, and every generation fears for what structures and virtues have been spared so far. Regarding structures, Marx and Engels memorably noted in 1848: "All that is solid melts into the air, all that is holy is profaned."[31] Then came Nietzsche, and in our days, Daniel Bell, Christopher Lasch, and a host of others who deplored the softening of vigorous morals.

Smith's and Hume's apprehension were followed in 1896 by William James's call for "The Moral Equivalent of War" as a setting where in peaceful times the "military ideals of hardihood and discipline would be wrought into the growing fiber of the people."[32] The details of the process, as James imagined them, strike us today as strange, not to say bizarre, and are worth quoting because they show how intractable the predicament of courage had really become.

> To coal and iron mines, to freight trains, to fishing fleets in December, to dishwashing, clothes-washing, and window-washing, to

> road-building and tunnel-making, to foundries and stoke-holes, and to the frames of skyscrapers, would our gilded youths be drafted off, according to their choice, to get the childishness knocked out of them, and to come back into society with healthier sympathies and soberer ideas. They would have paid their blood-tax, done their own part in the immemorial human warfare against nature; they would tread the earth more proudly, the women would value them more highly, they would be better fathers and teachers of the following generation.[33]

These words might as well have been written three hundred years ago for all the distance in situation and sentiment that separates us from them. Mining is a declining industry, freight trains have been passed by trucks and planes, steel furnaces have all but disappeared from the United States. Most important, we would not think of restricting whatever moral toughening to young men, nor would we confess, much less give praise, to "the immemorial human warfare against nature." We have crossed a cultural divide from the modern to the postmodern era.

Or have we? The conservative British philosopher Roger Scruton a few years ago was appalled by the lack of courage the Allies showed in the Kosovo conflict. Risk was for him the wellspring of moral vigor, and so he thought it was wrong for governments to promote "moral obesity by reducing risk in activities, consumer products, and employment."[34] Such a proposal would lead us from the bizarreness of James's untimely proposals to the bizarreness of wanton risk, maiming, and death. Imagine a world without lifeguards, seat belts, and safety goggles.

Miller's lament about postmodern courage exhibits a similar concern, though much more thoughtfully and with engaging ruefulness:

> There has been a recent spate of books and movies that look with great nostalgia on World War II, written by or directed by those who did not fight, who now in their middle age, when it is very safe for them to indulge this kind of wistfulness, think it vaguely amiss that they missed out on war. Most of my social class in the United States (myself included) bought substitutes for the only war we were eligible to fight in and would no doubt do so again. So when in middle age I come at last to believe that a nation builds up a moral treasury of

merit by the sacrifices of its people in war and I begin to worry, like the ancient moralists did, that we grow fat, lazy, and contemptible amidst our plenty, I don't have a leg to stand on to make that claim. My father could; he fought; but he is too wise to make it.[35]

We seem forced to conclude that courage has become superfluous. Would you teach your children how to handle an ornery gelding when you know they will never lay a hand on a horse? Or would you teach them that skill so they would be good at something else, say handling ornery people? But would it not be more helpful to teach them directly how to deal with difficult persons? So with courage. If learning physical courage is good for something else, why not learn the specific skill that is needed for that something? Yet with the passing of courage a m agnificent manifestation of what it is to be human would disappear as well. There is a principle of symmetry between reality and humanity that surfaces here. Great persons require great contexts. The principle of symmetry underscores another point. We have allowed to pass away not just a virtue but also a kind of reality.

The American paradigm of the receding challenge of reality is the frontier. Frederick Jackson Turner began his 1893 essay on "The Significance of the Frontier in American History" this way: "In a recent bulletin of the Superintendent of the Census for 1890 appear these significant words: 'Up to and including 1880 the country had a frontier of settlement, but at present the unsettled area has been so broken into by isolated bodies of settlement that there can hardly be said to be a frontier line. In the discussion of its extent, its western movement, etc., it can not, therefore, any longer have a place in the census reports.'"[36] As for the role the frontier had in shaping the character of this country, Turner's thesis was the following: "The peculiarity of American institutions is the fact that they have been compelled to adapt themselves to the changes of an expanding people—to the changes involved in crossing a continent, in winning a wilderness, and in developing at each area of this progress out of the primitive economic and political conditions of the frontier into the complexity of city life."[37]

Turner's thesis has come under fire for its thrust and its details. It certainly deserves criticism for its condescending and unrealistic portrayal of Native American culture. But whatever the prejudices and cruelties at the frontier, there is no doubt that the white people's move into Indian country has provoked courage among the pioneers and settlers even as the challenges defeated and crushed more than a few of those who went west. And whether or not the frontier has put a significant imprint on American institutions, it left us with the recollection of traits that we associate with frontier life and that Turner summarized thus: "The result is that to the frontier the American intellect owes its striking characteristics. That coarseness and strength combined with acuteness and inquisitiveness; that practical, inventive turn of mind, quick to find expedients; that masterful grasp of material things, lacking in the artistic but powerful to effect great ends; that restless, nervous energy; that dominant individualism, working for good and for evil, and withal that buoyancy and exuberance which comes with freedom—these are traits of the frontier, or traits called out elsewhere because of the existence of the frontier."[38] It's a fine description of American resourcefulness, of what we have been at our best, individualism excepted. In any case, the significance of the frontier for the virtue of courage is that it illustrates how irrevocably the setting has slipped away that once provoked daring and endurance as a matter of course.

Friendship

We think of friendship as a relationship rather than a virtue. But a virtue it was for Aristotle, and he devoted two of the ten books of the *Nicomachean Ethics* to the discussion of friendship. As a moral virtue it was for him a skill that had to be acquired and maintained. Its object was of course a relation, in fact the most important relation between two persons. In our culture, marriage is the exemplar of that relation. There is a fine asymmetry between friendship and marriage in Aristotle compared to the way it is viewed in contemporary America. Aristotle thought of the most important bond as normally obtaining between two men; he was unsure whether there could be such between a husband and a wife.[39] We think of that bond as typically joining a husband and a wife and are unsure whether we should recognize it as holding between two men. I certainly think we should, and we should honor it also when it joins two women, a case Aristotle did not even consider.

Although Aristotle was not all that explicit about the conditions the best kind of friendship has to meet, we can confidently list these four: First, the two friends have to be one another's equals; second, they must be devoted to one another's moral advancement; third, they must share a concern for some good thing; and fourth, they must enjoy each other's company.[40] Parents to this day test for these conditions when a son or daughter gives indications of a serious relationship. In the age of self-determination, they have learned to do this gently and obliquely, but they will raise questions like these: "So what kind of person is he? What does he like to do? What does he do for a living? What do you do together? Do you really like him? Is he good to you? How tall is he?" These inquiries are attempts to get some insight as to the first three conditions. Since ours is still the age of romantic love, parents can take it for granted that the fourth condition is met, at least as long as their offspring is young.

The model of romantic love is an unstable amalgam of freedom and fate. You are free to pick whom you want. Yet when you meet your beloved, you expect the fateful lightning of recognition and a thunder of overwhelming emotion, followed by conscientious courtship. You can't stage a thunderstorm, you can only hope and get ready for it. That's the traditional picture, and we still get it in the movies. But people rarely get it in reality. Hegel (1770–1831), who thought he had figured out the workings of reason, reasonably concluded that parental encouragement or some other rational decision should be the beginning and basis of marriage, and inclination and emotion should follow.[41] American pragmatism, however, evolved a seemingly practical alternative to fate and chance—the practice of dating. It is an officially sanctioned shopping for the best possible match though the initiative, as in courting, has remained mostly male. Beginning in the 1960s the trials that had once reached from socializing to "necking" and "petting" were further expanded to include sex and living together. But in spite of the parents' careful probing and the youngsters' diligent trials, almost half of all marriages fail and end in divorce.[42] Some marriages are destructive and should end in divorce, and there seem to be couples who can end marriages amicably and frequently. So sometimes divorce is not a bad thing, and at other times it does not appear to be such. Yet, as we all know from experience or close observation, most divorces are trying and terrible.[43] The pain of divorce is the tribute freedom pays to commitment. Most of us still look for a steadfast and abiding bond with the beloved. Why we do and how an enduring marriage rewards a couple are delicate topics. They invite smugness on the part of the happily and enduringly married and resentment on the part of the unlucky.

Since most of us still hope for that lasting commitment, even after a first or second divorce, we should ask ourselves why. Charles Taylor, who is in a position to know, remarked once that "there is a depth to love conferred by time."[44] There can be continuity and consolation in a life lived together. It provides security and context for children. The view into the past is open and unembarrassed. But all this is so because two persons have become or meant to become one. Lesser friendships come and go since they are more like attachments that allow for detachment. To be torn, however, from one's larger whole is painful and leaves scars, and those who marry and divorce without scarring have learned to avoid pain, but they have also unlearned to love.

Just as wisdom has suffered from the disappearance of the cosmic order and courage from the lightness of being, so marriage, our highest kind of friendship, has suffered from affluence. In premodern times, economics as much as ethics counseled couples to value their marriage. In England, just before the Industrial Revolution, to take one example, marriage was for ordinary folk the only route to relative independence. The alternative to marriage was to hire on as a servant to a family, either joining the large staff of a rich or noble household or being the sole servant of a middle-class family. A married couple, to the contrary, moved into its own house and ran its own farm or shop. Marriage, of course, was just one of the conditions of independence. The other was the minimal wealth of a farm or business, whether acquired or inherited. Having attained a measure of independence, wife and husband depended on each other to maintain or increase their wealth. The loss by death of one spouse jeopardized or ruined the economic base of the other. A deserting spouse exposed the remaining partner to a like misery and had only servitude for the normal prospect. Under such conditions, fidelity was well advised.[45]

If the pain that divorce causes spouses and children worries us, we have to think about ways of building again a context that is more conducive to the stability of marriage. Constraints that come into play once a marriage is already on the rocks are senseless and cruel and strike us as artificial. We have to discover the ground on which to build structures that naturally and continuously support fidelity and fortitude.

Economy

Economy, as I use the term, is the art and virtue of householding. Before there can be a household, there needs to be a house. To establish a shelter for the household is a fundamental feature of human culture. But what is the central ethical issue in building and dwelling? We can take our cue from Aristotle who was one of the first to discuss economy and did so in his *Politics*. In his treatise *On the Soul,* he framed two propositions that point us in the right direction.

Aristotle's first proposition says: "The soul is the form of the body"; the second says: "The soul is somehow everything."[46] Together these two principles provide a fair definition of the human condition. The vital force of a human being has a material center and a potentially all-encompassing comprehension of reality. The material focal point is first of all the human body, but then also the shelter that houses body and soul. As Kent Bloomer and Charles Moore have it, "at its beginning all architecture derived from this body-centered sense of space and place."[47] The cluster of habitats, the village, is one of the typical ways ancient human cultures marked their place in the world. So to mark and occupy a focal area of nearness is inhabitation. Here in Montana, at the edge of the northern Great Plains, such villages once consisted of the tipis whose inhabitants constituted a band.

The Europeans who invaded this area in the second half of the nineteenth century had little understanding and less appreciation of the ways the Blackfeet inhabited this region. What they did have in common with the Blackfeet was the desire to inhabit the land. Wanting to possess and command the land you live on and the house you live in is a powerful drive. It drove families across the Atlantic in miserable conditions and across the Great Plains under harrowing circumstances. Homesteading became the great American gesture—claiming a piece of land and improving on it. Claiming was always taking away, and improving at times was destruction. With this sorrowful proviso, the heroism of the settlers should not be dismissed. Once on their homestead, they piled up a shack from sod, hitched the mule to the plow, raised their chickens, hoped for rain, and were as often crushed as rewarded by their challenges.

The contemporary echo of this heroism is the desire to own your home. An astonishing two-thirds of American households do so.[48] Until the middle of the last century, moreover, the standard family residence exhibited a fine moral order. Kent Bloomer and Charles Moore published a paean to the typical house when what they praised was already being undermined.[49] They pointed to the public and formal front of the house with a respect-inspiring lawn and a stately entrance as distinguished from the enclosed and informal backyard. The rooms inside followed the distinction. On the public side you found the official living and dining rooms, toward the back the kitchen and sleeping quarters. At the center of the house was the "hearth (like a heart)," they said and later added: "A favorite painting might go over a mantel on which especially prized objects are placed, and the family's best rug and fanciest furniture are generally nearby."[50]

The particular and paradigmatic form of the house in the first half of the twentieth century was, against the best efforts of architects, the bungalow. Its open structure was to invite its rural setting in. Inside, the use of local timber and stone in turn disclosed the country outside.[51] Similarly the ranch house in the second half of the century was to recall the wide-open spaces of the West. But the ranch house took a decided turn toward the spectatorial and the opaque, replacing the bungalow's porch with the picture window and natural materials with machine-made and prefabricated elements.[52]

In any case, the primary function of house and home was enclosure, sheltering the life and integrity of the family; in the best case, enclosure also told us what it was an enclosure from. Enclosure ideally is disclosure too. The bungalow, though often truthful in materials and artisanship, faked the disclosure of fields and woods since it was typically located in a suburb, and the ranch house, no matter its picture window, disclosed neither an actual prairie nor the timbers and rocks of the West.

The mortal malady of the house, however, infected enclosure rather than disclosure. When Bloomer and Moore published their book in 1977, the fireplace at the center of a home's inner space, "the heart and hearth of it," had already been replaced by the television set, and the privileged chamber of the center, the living room, had begun to shrink and was to disappear entirely in many cases to make room for the informal comforts and the television set of the family room.[53]

Television is the eight-hundred-pound gorilla of social reality and social science. It sits down wherever it wants, and it turns out to be sitting everywhere. Over 98 percent of all households have at least one set, half have three or more, 60 percent of children between eight and sixteen have one in their bedroom, and so do a quarter of the children under two.[54] People watch hours of it every day. Look at any social research that takes a broad view of contemporary American culture, and before long the same old gorilla comes into view: Television, more than anything else, has transformed American life in the last half century.[55] It has made us more sedentary, passive, and disconnected from one another, or as it is sometimes put uncharitably, television has made us fat, lazy, ignorant, and selfish. We have strayed far from the Aristotelian condition of vigorous bodily life and wide-ranging comprehension of the world.

Business has been pushing this development, to be sure, but consumers have been yielding too. The push, moreover, is part of a broad movement toward a more sedentary and superficial culture. It has taken us from the bungalow to the ranch house, it has swept television into the ranch house, and whenever everyday culture appeared to catch its breath and settle into a more solid way of life, some technological innovation has stirred things up again and expanded the reign of passive consumption.

In the 1960s, it seemed that television viewing had peaked and settled on an average of ten hours a week. Then within a decade it grew by half. The reason was the appearance of a more captivating form of the medium—colored television.[56] Today, in the early twenty-first century, huge digital high-definition screens will again sharpen the attractiveness of television. This development shows once more how obtuse we are about Churchill's principle. Government and industry are consciously and vigorously pushing digital television and yet are blissfully ignorant about the way we will be shaped by what we are shaping. The television industry was at first reluctant to invest in the new technology. So Congress took the lead and decreed that "the current broadcast television service must eventually convert completely to digital," and it set a target date of December 31,

2006.[57] Once the inevitable was obvious, industry did its part to spread the availability of digital high-definition television (HDTV). In December of 2002, the Consumer Electronics Association and the National Cable and Telecommunications Associations agreed on rules and standards that make reception of HDTV programs via cable a matter of plug-and-play.[58]

But what is digital television? The Federal Communications Commission has the answer: "Digital Television (DTV) is a new type of broadcasting technology that will transform television. Because DTV is delivered digitally, it allows for the delivery of a signal virtually free of interference. DTV broadcasters will be able to offer television with movie-quality pictures and Dolby digital surround sound, along with a variety of other enhancements."[59] The FCC, addressing the American public, says it "is working to make sure the nation's communications systems are working seamlessly and competitively in your best interest."[60] But is it? It's not wrong to say that DTV will transform and enhance television. It will enhance it technically and yet make it more injurious culturally. The solution, of course, is not prohibition, but there should at least be a kind of warning such as the Surgeon General has issued about cigarettes and alcohol. Better yet, there should be a call for constructive (and noncoercive) counterforces.

Once again consumers are yielding to what industry and government are pushing. The "movie-quality pictures and Dolby digital surround sound" that the FCC has promised, both from television and DVDs, are given a place of honor in home theaters. As was the case with early television, there is an illusion, well chronicled by Katie Hafner, of communal celebration centered on the big, high-resolution screens. "People really congregate around it," Hafner quotes a designer and installer of home theater systems. "It's what people do now. It's an American temple and the screen is the altar."[61]

There are two problems with this. Home theaters depopulate the larger and more important congregations of movie theaters and sports venues. A community thrives when people regularly venture out into it, meet people beyond their narrow circle of friends, and begin to feel responsible for their city and their fellow citizens. The other problem is the short half-life of excitement generated by technological innovations. The bright new screens will become old. Moving solemnly into the home theater will seem awkward and artificial. And why bother the Joneses with an invitation to watch *L.A. Confidential* at seven when they own a screen just as big and bright, may not be ready by seven, might prefer musicals to film noir, and may just not feel like leaving their house in any case.

We can guess that the big screen and the big sound will be integrated into the family room, grandiloquently called "the great room." Viewing patterns will be much the same as before with one exception—there will be more viewing than prior to DTV, and Internet browsing along with DVD watching will take, at least proportionally, time from television.

What comes after DTV? Mark Weiser saw the progress of computer technology as a logical progression. It goes from one computer serving many persons, via one computer serving one person to many computers serving one person.[62] Or more technically put, the era of the mainframe computer was followed by the era of the personal computer, and what awaits us is the era of "ubiquitous computing." This phrase, say Weiser and John Seely Brown, will be "characterized by deeply embedding computation in the world."[63] A favorite instance of this development is the single-family house that, once it has computation deeply embedded in it, becomes a "smart" or "intelligent" house. You have seen it sketched and greeted with glad cries in the media. Still, let me remind you of some of its virtues. "Over the next twenty years," says Weiser,

> computers will inhabit the most trivial things: clothes labels (to track washing), coffee cups (to alert cleaning staff to moldy cups), light switches (to save energy if no one is in the room), and pencils (to digitize everything we draw) . . .
>
> . . . the kind of tune the computer plays to wake me up will tell me something about my first few appointments of the day: A quick-urgent [*sic*] tune: 9 am important meeting. Quiet, reflective music: nothing until noon . . .
>
> . . . my see-through display and picture window will show me the traces of the neighborhood as faintly glowing trails: purple for cats, red for dogs, green for people, other colors as I request.[64]

Here are some examples that Weiser and Brown produced together: "Clocks that find out the correct time after a power failure, microwave ovens that download new recipes, kids toys that are ever refreshed with new software and vocabularies, paint that cleans off dust and notifies of intruders, walls that selectively dampen sounds, are just a few possibilities."[65] A few additional classics: A system that "lets you see who is at the door and talk with them via a video cell phone even when you're not at home," in your bedroom "an electronic health checker that will monitor the user's health and can also be programmed to send data to health professionals," and a "voice memo panel on the refrigerator."[66] A house that "after it scans your retina on the porch, unlocks the door for you. Once inside the lights come up, the blinds open and your favorite aria filters through the speakers."[67] And then there is the classic among classics—the refrigerator that keeps track of the quantity and quality of your milk and notifies the milkman as needed.

Bill Gates's house, not surprisingly, is smart already. When you are his guest, you will be given an electronic pin for identification, and these will be your rewards:

> When it's dark outside, the pin will cause a moving zone of light to accompany you through the house. Unoccupied rooms will be unlit. As you walk down a hallway, you might not notice the lights ahead of you gradually coming up to full brightness and the lights behind you fading. Music will move with you, too. It will seem to be everywhere, although, in fact, other people in the house will be hearing entirely different music or nothing at all. A movie or the news will be able to follow you around the house, too. If you get a phone call, only the handset nearest you will ring. . . .
>
> If you're planning to visit Hong Kong soon, you might ask the screen in your room to show you pictures of the city. It will seem to you as if the photographs are displayed everywhere, although actually the images will materialize on the walls of the rooms just before you walk in and vanish after you leave.[68]

I find the aimlessness, banality, and unreality of these scenarios overwhelming. At the same time I must stress that some of these technologies make sense once a sensible function has been specified for them, for example, that of helping elderly or disabled people to gain a measure of independence and security or an energy-saving function.[69] But when ubiquitous computing is presented as a new kind of environment that sponsors a new style of life, the tediousness and triviality of concrete examples is dispiriting.

One way of concealing these embarrassments is to summarize them with a grand gesture. This is what William Mitchell, Dean of MIT's School of Architecture and Planning, does. Mitchell thinks that information technology will dissolve and reconstitute the very architecture of bricks and mortar: "Increasingly the architectures of physical space and cyberspace—of the specifically situated body and its electronic extensions—are superimposed, inter-twined, and hybridized in complex ways."[70] Another way out is to recognize what ails the proponents of ubiquitous computing—a loss of nerve—and to let cynicism step in where enthusiasm has failed. Here is my modest proposal: Replace the windows of a house or apartment with large high-definition electronic screens, and inconspicuously embedded in the screens let there be heaters, cooling systems, blowers, and speakers. Let the screens be programmed so that they display any view you like, and emit any sounds you desire and any weather you please. Say you live in Detroit. You could then request to be awoken by the sight, the sounds, and the balmy trade winds of Hawaii. In fact you could have the course of the entire day's twenty-four hours follow a Hawaiian pattern. And there is more. With appropriate Web cameras in place and an eight-hour lag, you could spend your entire domestic life in Munich's Schwabing district, with the very weather, the people in the streets, the rumble of BMWs that actually took place in Schwabing eight hours ago. And let me add that wherever you may see problems of feasibility in my proposal, I see grist for the eager mills of information technology.

There is just one problem with so living in Schwabing—when you leave your house or apartment, you step from disarming Schwabing into unrelenting Detroit. But that step, ironically, is a moral obligation rather than a physical necessity, and since it is something we should rather than must do, we can refuse to do it. You do so by taking the elevator down to the garage, getting into your air-conditioned car, and suffused with classical music you glide to your downtown office garage to take the elevator to your office high above the grime and grimness of Detroit. You can now drive from Detroit to Yellowstone Park while narrowing your vision of the continent to the clues a soft GPS-guided voice gives you as you drive along while the children in the back seat of the van watch cartoons on the built-in electronic screen.

The human condition that corresponds to this state of affairs is the person reduced to a dimensionless source of free-floating desires. This is the endpoint of the development that began with the Aristotelian person, an embodied and sheltered human being whose crucial faculty was not desire but reason, what Aristotle called the *logikon,* literally the faculty of gathering the world in a coherent vision. But just as television has diffused our vision of the world, so the smart house will disable our engagement with the home. It will so indulge our laziness and anticipate our desires that we will slide from housekeeping to being kept by our house.

Is it possible not to yield and to reassert economy, to take responsibility for the shape of our houses? Consider television once more. What's wrong with having your two-year-old watch the video of a canoe "paddling around an idyllic northern lake full of wildlife" as Lynn Neary did with her daughter? You need to understand, says Neary, "how insidious the process is. You start with the 'Einstein' videos. Next, it's PBS. It's all educational, right? But then the kids get restless, and in a weak moment, you start to channel surf. Before you know it, your child is addicted to 'SpongeBob SquarePants' and no amount of self-delusion can convince you that 'SpongeBob' is educational."[71]

It's the sheer availability of comforts and pleasures that appears to distract us from what we really want and that robs us of the command over our surroundings. In a jewel of insight, Mark Patinkin has us watch him as he sits down at his computer to write a novel. With "history's most efficient writing tool" at his disposal, he feels at a great advantage over Shakespeare and his pen.[72] While pondering his lead sentence, Patinkin decides to check his e-mail—only a click away. Having gone through his mail and poised again to start his novel, he wonders about tomorrow's weather, the weather where his parents live, . . . Back to Microsoft Word and the novel. Incidentally, what sorts of novels are selling today? On to Amazon.com, the Segway scooter advertised on Amazon . . . back to the novel . . . on to the stock market . . . back to the novel . . . return to e-mail. In the end, Patinkin envies Shakespeare the simplicity of the pen.

Distraction is a widespread problem, as Katie Hafner has found.[73] It emanates from the information that overwhelms and suffocates our native curiosity. In a premodern setting, where news was relatively scarce and precious, being curious was a requirement of prospering and even surviving. "Today," Hafner reports, "there is a universe of diversions to buy, hear, watch, and forward which makes focusing on a task all the more challenging." She concludes her report with the plaint of Peter Hecker, a corporate lawyer: "'Deep thought for a half hour? Boy, that's hard,' Mr. Hecker said. 'Does anyone ever really have deep thoughts for half an hour anymore?'"[74] What can we do? Here is Neary's answer: "Now I know a lot of you are saying, 'Just turn the thing off,' and you're right. You're absolutely right, but these comments are not addressed to the parents who have the moral fortitude and good sense to do just that. No. I'm talking to the rest of you who think you have it in your power to hold back the cultural onslaught that starts with those cute little baby videos. Take it from one who actually knows half the lyrics on 'The Lizzie McGuire Movie' soundtrack. You can't. It's a Pandora's Box, and if you don't want to deal with the demons inside, don't turn it on in the first place."[75] Neary recognizes the force of fortitude, but she overlooks economy. Lack of fortitude *and* of economy is fatal. If the set remains across from the couch and the remote remains on the coffee table, television will usually be turned on. But when we realize the importance and responsibility of ordering the household, we will either throw out the television sets simply or we will keep one in a place such that a conscious decision needs to be made and actual steps have to be taken to turn the set on and to watch it—a place such as the exercise room on the third floor, an out-of-the-way bedroom, or some place in the basement.

Students are discovering through experience what economy could have taught them right away. Electronic textbooks are toxic to concentration.[76] Churchill's principle tells you to reach for the seclusion of a real book. And as for the plaintive victims of a computer's distraction, here is a remedy drawn from Churchill's principle: Get a notebook computer, put it in your desk drawer, write your important stuff in longhand, and more than half an hour of deep thought can be yours.

Grace

The good life requires more than a shrewd arrangement of the household. Life needs to be full of grace, whether secular or religious. It can suffuse a mountain lake, and it can be the presence of the divine. Art is the realm where once the several appearances of grace were one.

An occasion where grace was resplendent in its many ways was the procession of the Panatheneia in ancient Athens. It took place every four years on the birthday of the goddess Athena and wound its way from the Dipylon city gate, across the Agora, up to the Acropolis. We have a vivid picture of the procession from a frieze at the top of the inner body (the cella) of the Parthenon, the greatest temple on the Acropolis, sacred to Athena. (The frieze is now in the British Museum in London.) There we see handsome youths on spirited horses, noble chariots, venerable elders, festive musicians, stout cattle, decorous women, solemn officials, and, reclining serenely, the Olympian gods and goddesses, larger by a third than the humans and in easy company with the mortals, too easy company, perhaps, for the Peoples of the Book.[77]

How close this depiction was to reality is controversial. But there is no doubt that art and religion were still one then as they were with the Blackfoot Indians up into the first half of the nineteenth century. The sun dance was the Blackfoot counterpart to the Panatheneia. Blackfoot art was resplendent at the festival in the finely tanned and decorated buckskin shirts, in the commanding eagle feather headdresses, the fearsome grizzly claw necklaces, the colorful bison hide lodges, and the mysterious medicine bundles.[78]

The unity of art and religion has been falling apart in the West for two hundred years at least and along many fault lines. The line between religion and art, as we think of art now, seems to be the crucial one, but it is not. Great art still moves and inspires us in quasi-religious ways, and religion, apart from its charitable works, most impresses us when it has its epiphany in art.

The really ruinous division is between high art and popular art. High-mindedness has gone one way, vitality another. There have been occasions where high and popular art became one and produced captivating works. But they were not to prevail. Claudia Roth Pierpont has described a case in point: "In the fall of 1924, the return of Aaron Copland from his studies in Paris and the arrival of Louis Armstrong from the clubs of Chicago began to tear the ideal of an American synthesis apart. Serious music and jazz went their distinct ways—a composer's art and a performer's art; eternal verity and restless improvisation—leaving Gershwin in a no man's land that came to be known, with a curl of the lip, as 'pops.'"[79]

To preserve high art is our duty, no doubt. But what do the well-to-do subscribers take from the New York Philharmonic into their daily lives? What difference do the millions of visits to the Metropolitan Museum in New York make to the lives of the common visitors? Even if constructive answers to these questions were available, one could point out that not only the symphony subscribers but also the museum visitors presumably come from the upper middle classes and in any event constitute a tiny percentage of the American population. What about the countless masses? And doesn't just raising this question remind us that ordinary people are simply not interested in substantial art? If this is so, it hasn't always been that way. In the nineteenth and early twentieth century "simple folk," unskilled laborers and domestics, were often avid readers of Shakespeare, Milton, Coleridge, and similarly classic authors.[80] Evidently the shape of the common culture no longer makes ordinary people reach for *Julius Caesar, Lycidas,* or *The Rime of the Ancient Mariner*.

Popular art is much more pervasive in American culture. Film, television, and popular music permeate our lives and homes. Much of popular art is garbage—clichéd, formulaic, lurid, prurient, and exploitive. Still, it is often the only voice that for ordinary people says out loud and memorably what we fear, suffer, hope for, and desire. But, alas, the illuminations such art provides are brief and do not carry very far. They have nothing of the sense of order and consolation the faithful used to receive from religion when it was one with art.

CHAPTER ELEVEN

Political Virtues

Justice

Justice is a virtue that looms large in Aristotle's *Nicomachean Ethics*. For Aristotle, as for the Bible, justice in its fullest sense was the unity of all virtues. When the Hebrew Scriptures or the Gospels really wanted to praise a man, they called him just. Noah was a just man, so was John the Baptist, and so was Cornelius the Centurion.[1]

The justice that mostly concerned Aristotle and mostly occupies us today is more limited in one sense and wider in another. It is a political rather than a personal virtue, and it defines the terms on which we think it's fair to live together. John Rawls has given us the best theory of what those terms should be.[2] As Rawls reportedly knew, we in this country have drifted away from our best, from the Rawlsian, conception of justice in the more than thirty years since Rawls's theory was first published.[3] Not that the Constitution is in jeopardy or there is more crime in the streets. But the spirit of fairness and cooperation that would give life to Rawls's second principle of justice, the one that calls for solidarity with the poor and powerless, that spirit has been waning.

Fairness is a matter of civility more than legality, of civic-mindedness rather than law-abidingness. Robert Putnam has studied the decline of civility (he calls it "social capital").[4] Its fate, as graphed over the time of the twentieth century, appears as a trace we could call the Putnam curve. Civic-mindedness rose at the beginning of the twentieth century in reaction to the excesses of the Gilded Age and in the wake of the Progressive Era, was further spurred by the challenges of the First World War, reached a first high point around 1930, declined when people hunkered down in the Depression, rose once more in response to the Second World War, reached a second high point around 1960, and has been steadily declining ever since. Whether it's membership in voluntary associations, in unions, or in professional association, whether it's league bowling or philanthropic generosity, the same contour with minor variations emerges again and again.[5]

The surge of civic-mindedness during and after the Second World War has been credited to "the Greatest Generation," the one that rallied to fight totalitarianism, set rivalries aside to produce the weapons of war, and met the enemy on the battlefield; the one that established democracy in West Germany and Japan and helped to rebuild Europe; the one that sent the GIs to college and built a prosperous economy; the generation that did everything right, one thing excepted—it failed to transmit its culture to the next generation.

Putnam has shown that most of the decline in civic engagement is generational. It is not the case that "the Greatest Generation" gradually became more self-centered and materialistic. No, those people remained as engaged and cooperative as ever. But the succeeding generations more and more lost the virtues of their parents.[6] Where did the parents go wrong? Were they careless, mistaken, or irresponsible in educating their children? Or were they oblivious to Churchill's principle and failed to realize how the world they had shaped would shape their children?

Putnam's well-crafted book suddenly and sharply illuminates a development that we have known of dimly and in parts. But the relative invisibility of civic decline could also indicate that the decline is not as calamitous as Putnam thinks it is, or that the calamity in question is of a new and very different color, so different, perhaps, that the greatest generation was blind to it. Putnam warns that the loss of civic-mindedness endangers our welfare, our safety, our prosperity, and our health and happiness.[7] There is little evidence that overall this is in fact happening. It might be that civic-mindedness is simply becoming dispensable, that social machineries are beginning to provide the stability and affluence that once required voluntary associations, and that screens large and small are more and more replacing the pleasures that once we gave and received from each other in bowling leagues and at dinner tables. Justice in the warmer and thicker sense of solidarity and graciousness would then grow ever fainter, and few would notice.

My hunch is that the cocoon of stability and affluence that the culture of technology has spun around most of us has muffled the claims of solidarity with the poor and powerless among us. This muffled condition has produced the shameful indifference to misery that is so starkly revealed through the lens of monetary utilitarianism. What is worse, the claims of justice in the sense of fairness and inclusion have no national boundaries. We seem to be insensible to the global misery that we could alleviate without suffering a significant loss in comfort. At the level of ordinary people and voters—the crucial layer of democracy—this stinginess is not the result of a conscious and informed decision. People assume that foreign aid comes to 25 percent of the federal budget, and they would like to see it reduced to something like 10 percent.[8]

Few people know, however, what the federal budget is (about $2.2 trillion in 2003) and that the share for foreign aid is about seven-tenths of a percent (about $16 billion in 2003), so that what people think is a "reduced" share, 10 percent, would be fifteen times what we actually give. Such is the clouded view that Americans typically have of the world; and as a consequence perhaps, their sense of fairness is clouded with ignorance and indifference. There is some evidence that, if people knew more, they would be more ready to help.[9] But what is cause and effect here? Does indifference breed ignorance or does ignorance abet indifference? Perhaps the world's economic and military center of gravity that is the United States draws our attention inward when it comes to compassion (and outward as regards economic and political power). At any rate, our indifference is an unhappy distinction among the wealthy nations. At the top of the list of generosity are the Scandinavian countries that contribute a share roughly ten times what ours is. We are at the bottom of the list.[10]

Sometimes the consequences of the imbalance between our power and our awareness go beyond neglect. In December of 1981, over 750 men, women, and children were killed in El Mozote in El Salvador. The atrocity was committed by a regime that our government had supported and encouraged. When the disaster was reported in the press, our government did its best to deny it or play it down. Had not Mark Danner given a searing account of the entire affair, it would have left no trace on our common memory. Even so, you could not say that *The Massacre at El Mozote* is the moral landmark in American foreign policy that it should have become.[11]

Stewardship

Something similar has happened to the virtue of stewardship for the environment. No doubt American society crossed an environmental divide sometime in the late 1960s and early 1970s. Let April 22, 1970, the first Earth Day, mark the crest of the divide. Some twenty-five years earlier, in 1954, a movie was made in Missoula, Montana, and the surrounding area. It was titled *Timberjack* and told the story of Tim Chipman who, on the death of his father, returns to Montana to claim his inheritance of timberlands. The evil Crofts Brunner, however, has designs on Chipman's lands and timber as well as on the beautiful Lynne, so generic a woman as not to need a last name. Brunner keeps Chipman from using the railroad needed to carry the logs to the mill. Some regulation, moreover, seems to prevent Chipman from cutting the old growth trees on his properties at this time. Chipman is nothing if not resourceful. His inquiry reveals that he can in fact harvest the trees now. He dams a creek so he can float his logs to the mill. The viewer is treated to the glories of man conquering nature, of timber jacks limbing and felling giant ponderosas, and of bursts of water carrying logs. Needless to say Chipman conquers Lynne as well.

From this side of the environmental divide, the account of *Timberjack* sounds like a parody. No one today would think of making such a film in earnest. Timber still gets cut around Missoula, but whoever does it or proposes to do it carries the burden of proving that it is economically necessary and environmentally tolerable. No dams get built anymore; on the contrary, the dam across the Clark Fork, just east of Missoula, will be taken down.

This epochal change is not just a matter of sentiments but also of the conditions on the ground and in the air—they have significantly improved since the late 1960s.[12] Membership in environmental organizations skyrocketed in the 1980s and has remained high.[13] But in environmental stewardship too, the animating force of the movement, has been fainting. Environmentalism is now largely a matter of paid professionals and check-writing members. It has been unable to mobilize the government in the struggle against global warming.

There is a parallel between the decline of justice and the decline of stewardship. In both cases the requirements of stability and security have been honored, but those of generosity and care have not. In the case of environmentalism, part of the blame rests with environmental philosophers. They are, for one, preoccupied with internal divisions and orthodoxies and have ignored calls for a more unified and pragmatic approach.[14] For another, they have by and large failed to recognize that a shift in our relation to the natural environment is needed, a shift from respect for nature to responsibility for nature. Respect for nature in the sense of preserving it and leaving it untouched is no longer an option anywhere. We have involved ourselves with every last piece of wilderness. Everything in heaven and on earth is now put in our care.

Of course you *can* and in many cases *should* respect what you have to care for. But it is a respect of engagement rather than through distance. As important, while once you could hope that there was a natural definition for the respect through distance, the boundaries of care and responsibility are now open if not also unstructured. We have to take care of Yellowstone Park and have rightly concluded that it was our responsibility to reintroduce the wolf into the ecosystem. But now we are responsible also for the buffer zone between the Park and pastures, and if responsible for the pastures, then also for the buffer between pastures and cities, and if responsible for Montana, then also for Alaska, and if for Alaska, then also for Russia, China, and however far our ability to make things better reaches.

It reaches into the heavens as well as across the earth. As Bill McKibben has shown in *The End of Nature* (1989), we have touched and changed the once unreachable, the climate and the weather.[15] We in this country, even when we do not reach across our borders, are endangering the ways and lives of everyone on the globe through our contribution to global warming and through our indifference when it comes to establishing and enforcing policies of reform.

Wealthy nations have the means to do the research and take the steps that good stewardship requires. As the "2005 Environmental Sustainability Index Report" suggests, there is a correlation between a country's standard of living and its stewardship. Among the rich nations, the Report says, "[t]he United Kingdom, Belgium, and the United States fall well below the regression line—indicating sub-par performance given their level of wealth."[16] As a society, we ignore both the newly quantitative and the newly qualitative dimensions of stewardship. The quantitative scope is now global. The qualitative challenge is, as Eric Higgs has shown, to envision the task of stewardship, as a creative enterprise, welding requirements of ecological fidelity, economic efficiency, and cultural sensitivity into a vision that fits a particular place.[17]

Design

While political virtues of justice and stewardship have lost their vigor, the virtue of taking moral responsibility for the built environment, the virtue I have been calling design, is still struggling to be born. We have taken fair responsibility for the physical soundness of the structures we inhabit, though the American Society of Civil Engineers in 2005 gave us no more than a D for the shape our infrastructure is in, and it called for an investment of $1.6 trillion over the next five years.[18]

Like justice and stewardship, the virtue of design transcends the boundaries of this country. Our infrastructure has also become a superstructure, a construction of the global atmosphere and the climate. The design of this superstructure is anything but sound, and atmospheric engineers would surely give us a grade of F and call for many trillions of dollars to render our structures atmospherically benign. Design, moreover, transcends its own boundaries and overlaps with justice and stewardship. The bad design of the atmospheric structure most injures the poor around the globe and will do so more and more. It is also bad stewardship and will upset and impoverish the environment.

Especially when it comes to the moral and cultural quality of our lives, the virtue of design has had a hard time asserting itself. Take the rise of radio-frequency-identification chips that everyone expects to be embedded some day in every piece of merchandise.[19] There are technical obstacles to be overcome, but once these problems have been solved and you too have a chip embedded under your skin, you will be able to take whatever from the shelves of a supermarket, put it in your cart, wheel the cart through some reading device, and be on your way. The electronic reader will register you and the merchandise, compute what you owe and debit the sum to your account—one more step toward Bill Gates's "friction-free capitalism."[20]

What sort of moral concerns does this development typically provoke? There are concerns about the violation of people's privacy when business can collect personal and intimate information and thieves or the government may be able to get their hands on it. There are concerns that the unions may oppose the loss of jobs.[21] But few critics seem to notice that the reduction of social friction is also the further fraying of the social fabric. Unplanned encounters with strangers are one of the ways in which the texture of society is kept from unraveling.[22] The answer is not to keep cashiers chained to their stations. The point is that if there were something like a recognized political virtue of design, we would take explicit account of this loss of public interaction and provide for counterbalancing engagements.

The rise of obesity has stirred up a debate that illustrates how scattered and confused we are about the moral significance of design. Physical health is an obvious concern, and yet the task of taking responsibility for obesity is met with hostility or bewilderment. Is the individual responsible or is it society? And if society, is government to blame, or business? What about the conditions that are conducive to obesity? Are they physical or social? There are some good answers to these questions, but they have remained disjointed and largely invisible to the public. One or another issue pops up in the media only to sink into oblivion, leaving barely a trace in the collective memory.

To begin with the facts of obesity, in 1960–62, 44.8 percent of adults (20–74 years of age) were overweight. By 1999–2002, the percentage had risen to 65.2 percent. As for obesity, 13.3 percent of adults were obese in 1960–62; 31.1 percent were obese by 1999–2002.[23] What about the causes and consequences of obesity? In 1990, poor diet and physical inactivity accounted for 14 percent of mortality in the United States, and 300,000 people died as a consequence. By 2000 the percentage had risen to 16.6 percent, and 400,000 people died because of poor diet and physical inactivity.[24] If the trend continues, "poor diet and physical inactivity may soon overtake tobacco as the leading cause of death," say the authors of a report on the "Actual Causes of Death in the United States, 2002."[25] What's worse, according to the editorial on the report, "[t]he estimates suggest that diet/physical activity patterns are now in fact likely greater contributors to mortality than tobacco is (and, in retrospect, probably were in 1990) and are most likely increasing in their impact."[26]

What has happened in the last ten, twenty, thirty, or forty years? Did Americans individually and inscrutably decide to eat more and more? Or did the social and physical design of American society change so as to make excessive eating increasingly the default behavior of ordinary Americans? Here again our society may be in the vanguard of technological progress, the first to pass through a stage of food commodification that other countries are only approaching.[27] Americans on average are more overweight (have a higher body mass index) than the French, Japanese, or Belgians.[28] Immigrants to the United States, whether white, Latino, or Asian, are less obese on arrival than native-born citizens. However, the longer an immigrant lives in this country the more likely the person is to approach American levels of obesity.[29] European countries and the global population are following in our footsteps.[30]

The wages of avant-garde commodification may be visible in the uniquely conflicted ways we deal with food, at least in comparison with the French, Belgians, and Japanese. Although we worry more about food and health than the citizens of those other countries do, we both eat more and enjoy it less.[31] "Americans," the authors of the comparative study note, "seem to have the worst of both worry worlds, the greater concern and the greater dissatisfaction. So the virtue which the Americans seek continues to elude them."[32]

This apparent paradox fits well, however, with the futile pursuit of pleasure noted before. A pleasure that gets detached from its context of engagement with a particular time, a special place, and a beloved community can become addictive, especially when a lot of technological sophistication is employed to refine the pleasure and make it easily available. "Context of engagement" is not a category social psychologists use in their questionnaires. But some of the prompts of the comparative study clearly probe that context. Consider these two:

> I would rather eat my favorite meal than watch my favorite television show.
>
> I have fond memories of family food occasions.[33]

On these and other items, dealing with the pleasure and centrality of food in one's life, Americans are less affirmative than citizens of other countries.[34] To Americans, a meal seems less of a focal occasion than to people in France, Belgium, and Japan, though evidently the traditional culture of the table is threatened in those countries as well.

The place of food and eating in our lives is a moral problem in the broad sense of morals that includes both culture and ethics, and if things are going badly from the moral point of view, the question of responsibility has to be faced. So it was in Congress in 2004, though politicians are even less likely than social scientists to worry about contexts of engagement with time, place, and community. Hence, the broad issue came to a narrow point of contention—the problem of obesity that we began with.

On March 11, 2004, the Republican majority in the House of Representatives beat back the attempt to blame (and to sue) the fast food industry for the injuries it has allegedly inflicted on the obese. Such lawsuits are surely the wrong way to better design. But the Republican rebuttal amounted to disavowing any common responsibility for the context of eating, a context called "the toxic environment" by Kelly Brownell and Katherine Battle Horgen.[35] The Republican legislators bravely called the "cheeseburger bill" the Personal Responsibility in Food Consumption Act, and James Sensenbrenner, chair of the House Judiciary Committee, explained the sense of the act thus: "This Bill says, 'Don't run off and file a lawsuit if you are fat.' It says, 'Look in the mirror because you're the one to blame.'"[36] To which Eleanor Holmes Norton, Delegate of Washington, DC in Congress, replied: "We're talking about a public health problem for which our government has not taken responsibility."[37]

While taking responsibility through our government (as certainly we should), we also need a sustained national conversation on the design of contemporary culture. What parts of the cultural framework should we particularly look at when it comes to obesity? There are immediate and homely issues, close to the fast food industry, such as "portion size, price, advertising, the availability of food and the number of choices presented."[38] There are less visible ones. One factor that seems to aggravate obesity of the poor is the lack of grocery stores with good produce departments in their particular neighborhood.[39] Other factors seem distant from the times and places where people buy and consume food. In the 1970s, the Nixon and Ford administrations changed agricultural regulations to promote production and bring down food prices.[40] A chain of causes and effects was the result.

First came an abundance of corn and falling prices. Next came the need to turn corn into something more concentrated and valuable. Lots of chicken and beef was one result. Another was the invention of a process that turned corn into high-fructose corn syrup.[41] All this resulted in more abundant, more enticing, and more affordable food. To prevent a drop in income due to falling food prices, the food industry offered supersized portions of food and drink. And finally it turned out that people, presented with cheaper food, sweeter taste, and larger portions, eat more than before, reliably and unthinkingly.

But not unfeelingly. Evolution has left us with a chink in our genetic armor that makes us deeply vulnerable to excessive sweets and carbohydrates. In a culture, moreover, where we mostly sit and connect with the world visually, eating is one of the few ways we can sink our teeth into tangible reality. The emptiness and despair obese people feel when food is denied them is hard to fathom for the genetically favored.

It was, however, not only the redesign of agriculture, relatively distant and concealed to most Americans, but as much the redesign of the immediate and obvious urban fabric that led to the sorrows of overweight and obesity. It was the shift from the traditional city to the suburb. The ranch house is made for viewing, for watching television inside and for watching the outside world through the picture window. Houses are too closed in on themselves and too far apart from one another to invite neighborly interaction, and the houses are altogether too far from commercial and community centers to encourage walking. Suburbs are made for sitting at home or sitting in a car, and the more distended a suburb, the more likely a suburbanite is to suffer from obesity or chronic illness.[42]

Robert Kirkman has an exceptionally clear conception of design as a moral virtue: "A particular way of building is good to the extent that it makes it easier for people to live well, to be good people, and to participate in community life; it is bad to the extent that it makes these things more difficult by harming individual well-being, fostering injustice, fragmenting communities, or undermining the conditions of its own continuation."[43] But he has also found that people are either outright hostile to the very idea or they use "comfortably vague terms" such as "green space," "progress," "family," and above all "quality of life" to avoid an incisive conversation about design.[44] When decisions have to be made, the narrowing of moral discourse we have observed before takes place, and a "proxy battle" ensues. "A proxy battle," Kirkman explains, "is a conflict over a concrete problem or decision that stands in the place of much broader debate over basic values."[45]

There is some hope of a turn for the better, however, and more than hope, in fact. In 1981 the new urbanism was born in the design of Seaside, Florida, laid out by Andres Duany and Elizabeth Plater-Zyberk. To make the town "walkable" was one of the chief goals, but this is just part of the overall design that has for its purpose to create or rather recreate a vital urbanity. The newly urban towns have narrow streets, a regular grid, retail and civic centers reachable within five minutes of walking, houses with porches and on small lots with cars and garages relegated to the alleys, a mixture of apartment buildings and houses of different sizes and cost, all of this intended to create a diverse, active, and communal citizenry. Design as a political virtue has achieved its most self-conscious and vigorous realization in the new urbanism.[46]

Walkable neighborhoods do apparently encourage greater physical activity and less obesity.[47] Surprisingly, perhaps, the physical activity that is simply part of getting around on one's own power in daily life can have a significant effect on one's weight and health as research done at the Mayo Clinic suggests.[48] An article about the study explains how the lead author, James Levine, has managed to increase the amount of his everyday physical activity. A picture shows him "using a treadmill as he works on the computer in his office—something he does for about 40 hours per week—the equivalent of about 40 miles" as the caption has it.

It's a technological fix for a cultural problem, and it betrays the loss of engaging contexts and focal centers that bedevils our attitudes to food and at a deeper level bedevils the prevailing habits of design.[49] The point is not somehow to engineer physical activity into a technologically streamlined setting, but so to design our private and public spaces that they invite us to appropriate and enjoy them through walking and playing with body and mind. Whether, as regards the public sphere, the new urbanism will give us that sort of environment, it is perhaps too early to say though there is some encouraging evidence.[50] It is chiefly qualitative evidence. Regarding quantity, the new urbanism may have asserted itself on the pages of the *New York Times* and the *New Yorker,* but on the ground it has covered less than 3 percent of the housing units built between 1990 and 2000 in the United States.[51]

The Importance of Practice

A practice is a regular way of doing things. Design and economy play large roles in discouraging or favoring practices. But we should not overburden Churchill's principle. Even in the most favorably shaped circumstances, it takes some resolve and dedication to establish a practice.

Take voting. Here again we Americans are doing poorly in the community of nations. From 1960 to 2000, voter turnout in presidential elections has declined from 63 to 51 percent of eligible voters, though it approached 60 percent in 2004.[52] Still, on a list of 172 countries, averaging turnout from 1945–2004, the U.S. ranks 139th.[53] The theories about why this is so are many, but one that was acted on had it that voting was too cumbersome an affair. The convenience of casting one's vote had to be improved. Thus voters were told they could pick up their ballots and mail them in or have the ballots mailed to them as early as fifty days before the election. Some twenty-five states provide this kind of convenience.[54]

As usual, the intentions were good—to increase voter turnout and particularly to help citizens who "are infirm, away at college or otherwise unable to go to the polls," as one chronicler has it.[55] As often, the grasp of how circumstances influence practices was slack or simply wrong. "The concept of Election Day is history," Brian Lundy, founder of Helping Americans Vote has said.[56] If so, it's a great loss. Absentee voting, first of all and ironically, is intrusive. Partisans can find out whether you have voted or not. They can take the ballot to your door along with pressure to vote for their candidate.

More important, the quiet festivity of Election Day is in jeopardy.

Democratic elections are wonderful events. Traditionally, the furor of campaigning has finally come to an end. We all stream to the polls, in trickles now, in floods later. We see our neighbors and thank the election volunteers. The most momentous political event moves on the wings of a dove, and once the collective verdict is in, we all accept it in good grace. It is hard to believe that without joining together on one particular day and that without pledging ourselves to mutual expectations and obligations, elections can prosper as a vigorous practice.[57] In fact, widening the scope of early and absentee voting has done little or worse to strengthen voter participation.[58]

The "broken windows" theory claims that physical disorder breeds bad attitudes, that failure to keep the physical design tidy leads to moral disorder. But in this sense the inverse is just as powerful: Bad attitudes make physical disorder look worse than it is. One study has found that, among neighborhoods that objectively are equally littered and neglected, more disorder is seen by residents of all races in those areas where poverty and racial minorities are prevalent.[59] Whence it follows that attitudes need to be changed as much as physical circumstances.

When and how individual, social, and physical circumstances conspire to beget a practice is hard to say.[60] The practices that are centered on the culture of the table have favorable social and tangible circumstances on their side. Although the living room is disappearing from the American home, the dining room is not. It has remained dear to our hearts.[61] The family dinner, moreover, has much to recommend it as a practice. Teenagers who regularly eat dinner with their parents are less likely to use drugs, suffer less from depression, are less prone to suicide, do better in school, and they eat more healthful food.[62] And yet more and more often we eat away from the home.[63]

It must be grace that bears a practice, the grace of a focal thing or the grace of a credible advocate. Amanda Hesser tells of a moment of grace: "Parents who have dinner with their children often see it as a moment of focus. Barbara Kramer, a full-time lawyer in Ann Arbor, Mich., who is nine months pregnant and has two daughters at home, said, 'I just kind of made a decision when the kids were little that we were going to have this one time of day when we sat down together and we didn't answer the phone and we didn't engage in any other activity except for talking to each other.'"[64] Resolve does not really come from nowhere. It responds to the force of focal things and practices. If they have touched us, it's our task to be their advocates, but not only advocates. We also need to be architects who are devoted to the design and economy of a setting that favors the good life.

PART THREE

Real Ethics

CHAPTER TWELVE

Recognizing Reality

The Weakness and the Competence of Ordinary People

Who are they, those ordinary people? The ordinary person sleeps too little, drives too much, shops at Wal-Mart, eats at McDonald's, uses a refrigerator and a microwave oven, watches television, has a cell phone, logs on to the Internet, admires movie stars, envies billionaires, would like to lose weight and make more money. Being ordinary is a condition and a matter of degree. Everyone is an ordinary person to some extent.

People differ in the ways they conform to the ordinary condition, both in extent and intensity. The most ordinary people allow the standard conditions of American life most fully to shape their lives. Girls who are beautiful and voluble and guys who are handsome and athletic become high school stars, and unless they are also smart and driven, high school glory is the high point of their life. It's waiting on tables and construction labor from then on. Youngsters who are intelligent and ambitious go after the training and skills that are needed to grab a major share of affluence. They live the grand life that waitresses and laborers envy. But these successful people too live within the strictures of the prevailing culture. It's just that they live at the top of the established culture and earn and consume a multiple of what the average person gets.

It would be unrealistic and uncharitable to blame people for their implication in the ordinary culture of affluence and consumption. We take the jobs we can get, and we avail ourselves of what is at hand.

Workers and consumers have a lot of choices, but they are all of the same kind. Whether you pick McDonald's or Wendy's, Burger King, Taco Bell, or any of the dozens of fast food outlets, the quality of life remains much the same. You can take a job as a waiter, a receptionist, a salesperson, a grounds keeper, a janitor, or a short-order cook, but there is not much creativity or responsibility in any of them. Fortitude and camaraderie lend those jobs what grace they have.

There is a real question, then, whether ordinary people are at all capable of changing their lives for the better. Although as laborers and consumers they may be constrained, not to say paralyzed, by contemporary culture, they are fully competent as citizens to discuss and determine the quality and shape of their lives. When put in a reflective mood by appropriate questions, people give valid and reliable assessments of their happiness, their hopes, and their standing in society. In at least some cases, well described by Andrew Feenberg, people have actively and creatively adopted a technology beyond its original engineering conception.[1] Exhibit number one is the Internet. It was conceived as a redundant and resilient military communications network that could absorb damage to large parts of its structure. From there it morphed into a communication tool for scientists, and finally lay people appropriated it for mail, socializing, shopping, information gathering, political activism, and more.

John Rawls was right then in saying that everyone is a moral agent and that part of being one is the ability to form a conception of the good life.[2] But it is evidently an ability that can go dormant and needs to be reawakened in a common conversation. The conversation, moreover, needs to be more than procedurally correct. We have to give it substance and content and give our conclusions a tangible design. In any case, the good society will not come about through the forceful imposition or the sly engineering of an elite. It needs the conversation and sanction of ordinary people.

Yet granted that they can talk about the good life, can they lead a life of wisdom, courage, friendship, and grace? To begin with the most doubtful virtue, grace, can ordinary people appreciate and practice the arts? There are enough instances to show that the culture of the table, the culture of the word, the culture of music, and the culture of the crafts were once the secure and admirable possessions of common folk, and high culture itself used to be inspired and nourished by folk culture and enjoyed by ordinary people.

Lacking large amounts of expensive foodstuffs, poor people, given half a chance, have been more resourceful and inventive than their betters in the dishes they brought to the table. Ordinary people have been accomplished tellers of tales, thoughtful writers of letters, and perceptive readers of literary classics.[3] In this country, jazz has been the most remarkable upwelling of great music from common circumstances. Finally, the furniture and utensils of the Shakers are sought out to this day for the grace and simplicity of their construction.

Not everyone was an expert cook and narrator and musician and artisan. But often people excelled at least in one of these arts or they had expert practitioners in their family or neighborhood. Culture was deeply and widely rooted in, for example, colonial New England, and remnants of the old ways are scattered throughout contemporary culture.[4] Now most of those things and practices have been commodified and sold back to the people who once owned them.

Commodification and Contemporary Culture

Contemporary culture has given us unprecedented liberty and prosperity. This needs to be stressed and has been rightly and frequently celebrated. But it also needs to be understood. The characteristic liberty of American society is real and personal, not political. To be sure, political liberty, freedom from oppression and for self-determination, is fundamental. Without it nothing much matters except survival. But political liberty, though it needs constant vigilance and protection, is well understood and has been the object of hopes and struggles for hundreds and thousands of years.

Real liberty is the freedom from the burdens of reality, from hunger, cold, disease, and confinement. Personal liberty is freedom from the demands and annoyances of persons. Real and personal liberation began in earnest with the Industrial Revolution and has since become a broad and diverse phenomenon, so much so that we may despair of getting a grip on it and taking responsibility for it. Yet if there is anything to the sense that contemporary American culture is not conducive to a life of excellence and a good society, we need to find some way of identifying the central driving force of our time.

Commodification goes a long way toward disclosing that force and its liabilities. It refers us to a crucial structure of contemporary society, namely, to the market, and it conveys a sense of moral censure. "In recent years," to take an instance Viviana Zelizer has mentioned, "economic prophets have frequently warned us against global commodification and the loss of the moral-emotional fiber it brings."[5] The problem with commodification is that, where it is precisely defined, it loses its crucial force of cultural and moral criticism, and, where intuitively it appears to point to a central moral issue, it fails to be precise. What we need is a precise and incisive moral conception of commodification.

Where does the concept of commodification come from? From Marx of course, more precisely from his analysis of *die Ware,* a term that in 1886 already was translated as *commodity*.[6] The precise and economic definition of commodification derives from the verb *to commodify*. To commodify something is to draw it from outside the market into the market so that it becomes available for sale and purchase when before commodification it was not. Marx did not have a word either for the verb or for the process of turning something into a commodity. These were coined in Anglo-American Marxist philosophy of the 1970s and then migrated into other languages.

Marx analyzed and deplored to various degrees the ways in which capitalism took the production of things such as wheat, shoes, and clothing out of the hands and circumstances of farmers, artisans, and householders, stripped them of their context of skills and persons, of exchanges and uses and made them into commodities—goods whose significance seemed to be reduced to their price.[7] Marx criticized and lamented with even greater chagrin the conversion of labor into something that was bought and sold under conditions typically favoring the capitalists and beggaring the workers. For Marx, commodification was exploitation.[8]

So it was here in the nineteenth century. Industrialization, the mechanics of commodification, was in the United States, as it was elsewhere, a disruptive and violent storm whose benefits were slow in coming to the lower classes. Again we are the forgetful heirs of a prosperity that was built on the backs of men and women who toiled for seventy or more hours a week under miserable conditions and for very low wages. Worse was child labor. At the turn from the nineteenth to the twentieth century, nearly two million children between ten and fifteen years old were exploited, crippled, or crushed by hard and sometimes dangerous work. It was only in the late 1930s that child labor was nationally and effectively outlawed.[9]

In the advanced industrial countries, exploitation has been reduced in some areas and has disappeared in others. Globally, of course, it remains a calamity and a scandal. At any rate, commodification as Marx thought of it (with exploitation subtracted) is no longer controversial. For us, it is a matter of course that iron, paper, wheat, clocks, linen, bootblack, silk, diamonds, and gold are commodities and that for employment there is a labor market.[10] If we have complaints about the ways work is sold and bought, they concern, if anything, the failure of the labor market to follow the law of supply and demand punctiliously.

Yet this does not mean that commodification is uncontested in contemporary discussions. There is broad agreement that certain goods, such as justice, should never be for sale. On other goods, however, opinions are divided. To get a sense of what sorts of goods are at issue, consider Michael Walzer's list of items subject to "blocked exchanges": (1) Human beings; (2) political power and influence; (3) criminal justice; (4) freedom of speech, press, religion, assembly; (5) marriage and procreation; (6) the right to leave the political community; (7) exemptions from military service, from jury duty, and from any other form of communally imposed work; (8) political offices; (9) basic welfare services like police protection or primary and secondary schooling (they "are purchasable only at the margins," Walzer adds); (10) desperate exchanges (trades of last resort); (11) prizes and honors of many sorts; (12) divine grace; (13) love and friendship; finally, (14) a long series of criminal sales.[11] To this wise and circumspect list we can add public goods such as clean air and clean water, safety from crime, basic health care, and public lands. It is apparent from Walzer's list that goods are drawn or in danger of being drawn into the market from two areas, the realm of public goods, education for example, and the realm of intimate goods, childbearing for example. The master argument against the commodification of public goods is that commodification leads to social injustice. If education is fully commodified, the children of the poor will get no education or inferior education. The chief argument against commodification of intimate goods is that commodification debases something that is dear or sacred to us. Love, offered for sale, is corrupted and cheapened.

In the United States, commodification of public goods has been advancing in the last twenty-five years. Basic health care is a public good in several advanced industrial countries. In the United States it is available only in the restricted sense that anyone has a right to walk into the emergency room of a nonprofit hospital and to be treated for urgent illness or injury. Attempts of the early Clinton administration to provide more substantial universal health care failed spectacularly. Meanwhile, for-profit medical institutions and health maintenance organizations have strengthened market principles to the detriment of the professional discretion of physicians. Less visible and more subtle commodification is taking place in the pharmaceutical industry and in surgery where the line between healing and enhancement is being blurred and crossed. As a consequence there is a troubling shift of resources away from healing the ailments of the poor to the morally dubious enhancement of the rich—one more indication that a moral conception of commodification is needed.

Education is a second public good that is being diminished by commodification. In higher education, public support has been declining sharply over the last quarter century. Thus a degree from a public university is no longer something that every qualified student has a right to; it has become an expensive commodity. The elite private universities have tried to refashion access in the image of a public good—it's accomplishment rather than money that entitles you to enroll. But this well-intentioned reform is being subverted by the commodification of prepping for the Ivies. It takes rich parents for a child to get into the prestigious high schools, to engage in glamorous extracurricular activities, and to take the Kaplan courses that produce top scores in the Scholastic Aptitude Tests, much as the Educational Testing Service may deny that its tests are sensitive to prepping.

David Noble sees a more insidious commodification of higher education in the rise of distance education.[12] It is not only access that in his view is commodified, it is the very thing, education, that is turned into a commodity. This kind of commodification, Noble contends, recapitulates the exploitation, de-skilling, and immiseration of the artisans that was effected by the Industrial Revolution. This time it is the professors whose work is converted by administrators and corporations into a digitized teaching module that can be sold over the Internet so that the expertise of professors is expropriated and most professors and the universities themselves become dispensable.

What Noble has predicted dolefully, Lewis Perelman and Peter Drucker have proclaimed gleefully—universities as we know them will become obsolete.[13] In reality, however, a division of clientele, circumstances, and value seems to be evolving. Certain kinds of students in particular circumstances do benefit from distance education in acquiring a degree of substantial if limited value.[14] Hence the commodification of higher education through information technology is mildly beneficial to social justice. It does broaden access to some degrees.

Commodification and Social Justice

Still, the wider commodifying trend in education is injurious to social justice, and it is part of a broad sweep, early and impressively analyzed by Fred Hirsch.[15] When the rich can buy a good education for their children, can find fine outdoor recreation in country clubs, and can be sure of their security in gated communities, and when the rhetoric of "taxation is theft" is widely advertised, public goods will inevitably suffer.

Not that the market and public goods are always opposed to one another. Clean air may be enhanced through a market in air pollution credits. Whether this is so is in important part an empirical question though moral questions always intrude. If a market in air pollution credits reduces aggregate pollution more quickly and at lesser cost than regulations would, and if at the same time some areas remain disproportionately polluted, is the market approach acceptable?

Turning now to the commodification that draws a thing or practice from the intimate sphere into the market, consider childbearing. Commercial surrogate mothers agree for a price, usually tens of thousands of dollars, to bear another couple's child, growing from the couple's fertilized egg or from the surrogate mother's egg fertilized by the man's sperm. The legal situation is complex. Some states in the United States put restrictions on surrogacy, some prohibit it, and some regulate it.[16] But is commercial surrogacy morally permissible in any case?

Some of the social justice concerns surface here again. Margaret Jane Radin fears that poor women, usually the providers of surrogacy, are being exploited and that children are being harmed.[17] Elizabeth Anderson similarly argues "that commercial surrogate contracts establish relations of domination over surrogate mothers that are inconsistent with their autonomy and with treating women with respect and consideration." Surrogate contracts, she adds, "engage commercial norms for valuing children which undermine norms of parental love that should govern our relations to them, thereby degrading children to the status of commodities."[18]

The opposing view has been argued, as we have seen above, by Richard Posner.[19] If surrogacy is fully commodified, as Posner proposes, it becomes subject to the law of contracts. If society has a legitimate interest in protecting childbearing from commodification, as Radin and Anderson argue, commercial surrogacy will be outlawed, the position I support. If moral issues do arise in commercial surrogacy, but do not rise to a state interest, then surrogacy is permissible as long as the surrogate mother is protected and possible abuses are outlawed. This is roughly Richard Arneson's position.[20]

Goods and services can also move in the opposite directions, from within the market into the public or the intimate sphere. In the first case, the movement is called socializing, and although these days the mere mention of the term is the death knell to any attempt at moving things that way, basic health care should certainly become a public good in this country. The name of the other movement is "do-it-yourself." The home-power movement is an interesting example. Practitioners get off the commercial grid and generate energy within their households.[21] Smallholding is a broader, if occasionally goofy, case of the same movement.[22]

At any rate, moral concerns about commodification are specific to the area from which goods are drawn into the market. Problems of social justice arise primarily when public goods are commodified. Issues of debasement are of concern when valued things and practices are pulled out of the sphere of intimacy. These are piecemeal moral theories of commodification. The first, moreover, the one that raises questions of social justice, is only incidentally a theory of commodification. There are, however, moral theories that are specifically about commodification and give a unified account of the moral complexion of commodification.

Two such theories are nearly total theories—one says that almost everything ought to be commodified, the other says that nothing should be commodified. A prominent example of the first is the Chicago school of law and economics of which Posner is a representative. Marx's theory might be taken to be an example of the second.[23] In this interpretation, there should be no markets at all. Allocation of resources is to be determined by a rational plan, distribution of goods is to be governed by rationing "to everyone according to his needs." But even in the communist countries, there was something like a market for distribution.[24] More important, industrial production, no matter its ownership, implies a kind of commodification, not economic commodification, of course—yet another symptom of the need for a moral conception of commodification.

There are also discriminative theories of commodification that give a unified account of the moral issues involved. They furnish a principled line that allows us to discriminate between what should be in the market and what should not.[25] Radin, though she avows a "pragmatic spirit" and forswears the "search for a grand theory," nevertheless adumbrates something like a principle—those goods and practices should remain outside of the market that are central to the flourishing of personhood.[26] As she cautiously puts it, "I am arguing that something important to humanity is lost if market rhetoric becomes (or is considered to be) the sole rhetoric of human affairs, excluding other kinds of understanding."[27] Elizabeth Anderson's discriminative theory too claims to be pluralist rather than unified. But there is nonetheless one reason why certain goods are to be kept out of the market—the norms of the market fail to express adequately "our rational attitudes toward people and other intrinsically valuable things."[28]

All the theories of commodification I have discussed so far are based on the economic conception of commodification and on the centrality of the market to that conception. On that basis they raise the moral question: What ought to be in the market? The three answers are: (1) Everything. (2) Nothing. (3) Something. Yet the basic assumption fails to give a crisp and clear account about cultural intuitions and moral misgivings that have clustered around commodification.

If commodification is invariably an instrument of exploitation and oppression, as the second answer has it, why is commodification spreading in societies that have a great deal of prosperity and equality? Proponents of the first answer have a ready explanation:

The market is in fact the social arrangement that maximizes liberty, prosperity, choice, and opportunities for self-realization. It's a position that is held not only by utilitarians but also by libertarians like Arneson: "The opportunity to choose among a wide array of goods for sale and among a wide array of employment opportunities, and hence among a wide variety of lives, is thought to be a good way to facilitate thoughtful and informed refining of one's preferences and values. In this way the market enhances individuality."[29] But both answers can't be right. A recent Marxist appraisal says disarmingly: "The extension of commodification is a contradictory process: demeaning and dehumanizing, but at the same time liberating and progressive."[30]

Is it possible to discriminate between dehumanizing and liberating commodification and to keep the former at bay as Radin and Anderson propose? For all the subtleties of their analyses, Radin and Anderson stay within the limits of distributive justice theories that furnish an ethics merely of blamelessness. As long as you respect the line between market and nonmarket goods and practices, you are free of blame. This is an important moral project, and I agree with many of its conclusions.

It seems counterintuitive, however, that moral concerns cleave as neatly into blamable and blameless ones as the discriminative theories imply.[31] Surely some of the blameless commodifications show traces of the moral opprobrium that altogether prohibits economic commodification of the relevant things and practices. Although the moral considerations in the blameless cases do not raise issues that require legal sanctions, they do call for a conception that is morally illuminating.

Moral Commodification

The problem, then, is not that the ethics of blamelessness is unimportant, but that is insufficient. Insufficient to do what? To explicate the intuition that there has been a "narrowing and flattening of our lives," as Charles Taylor has put it. "People have spoken," Taylor says, "of a loss of resonance, depth, or richness in our human surroundings."[32]

This diminishment of our lives has taken place within the realm of legitimate and blameless commodification and therefore requires an analysis that reaches into the blameless life and answers the question of the good life. The appropriate conception of commodification turns out to overlap but not to coincide with economic commodification. It is also more comprehensive and connects naturally with what comes before and after commodification, namely mechanization and consumption.

We can gain access to moral commodification by considering a process in eighteenth-century England that E. P. Thompson has described in a classic essay, "The Moral Economy of the English Crowd in the Eighteenth Century."[33] It concerns the ways in which wheat (or corn as it is called in England) was traded, milled, and baked into bread. In a somewhat simplified and idealized view of the pre-commodity moral economy, "[m]illers and—to a greater degree—bakers were considered as servants of the community, working not for a profit but for a fair allowance."[34]

There was of course a market for wheat, flour, and bread in the sense that money changed hands. But it was a market that was constrained by regulations imposed by a paternalist government. More important, the market was more than an economic institution as Thompson explains: "In eighteenth-century Britain or France (and in parts of Southern Italy or Haiti or rural India or Africa today) the market remained a social as well as an economic nexus. It was the place one-hundred-and-one social and personal transactions went on; where news was passed, rumour and gossip flew around, politics was (if ever) discussed in the inns or wine-shops round the market-square."[35]

Local ties inevitably went with social ties. In the paternalist arrangement, "marketing should be, so far as possible, *direct,* from the farmer to the consumer," Thompson says and later adds: "Corn should be consumed in the region in which it was grown, especially in times of scarcity."[36] Finally there were temporal ties. Wheat was tied to a certain time through the harvest: "Few folk rituals survived with such vigour to the end of the eighteenth century as all the paraphernalia of the harvest-home, with its charms and suppers, its fairs and festivals. Even in manufacturing areas the year still turned to the rhythm of the seasons and not to that of the banks."[37]

These bonds were broken as the free market, advocated by Adam Smith, moved in on the paternalist market and wheat became a commodity. Thompson notes the promise that propelled the free market: "The natural operation of supply and demand in the free market would maximize the satisfaction of all parties and establish the common good."[38] One result was that "[a]s the century advanced marketing procedures became less transparent, as the corn passed through the hands of a more complex network of intermediaries."[39] A more distant result was that "[m]arketing (or 'shopping') becomes in mature industrial society increasingly impersonal."[40] What is the moral of this story? Thompson sees it as follows: "The new economy entailed a demoralizing of the theory of trade and consumption no less far-reaching than the more widely debated dissolution of restrictions upon usury. By 'de-moralizing' it is not suggested that Smith and his colleagues were immoral or were unconcerned for the public good. It is meant, rather, that the new political economy was disinfected of intrusive moral imperatives. The old pamphleteers were moralists first and economists second. In the new economic theory questions as to the moral polity of marketing do not enter, unless as a preamble and peroration."[41]

This conclusion falls far short of the richness and insights of its premises. As a Marxist, Thompson was unwilling to accept the moral promise of the free market, namely that it would "maximize the satisfaction of all parties and establish the common good." As a historian, he had to acknowledge that "[i]n some respects [Adam] Smith's model conformed more closely to eighteenth-century realities than did the paternalist; and in symmetry and scope of intellectual construction it was superior."[42] Unwilling to accept the everything and nothing positions on economic commodification, Thompson saw no other alternative than to adopt the middle position that makes peace with moderate and morally blameless commodification.

Yet both the lineaments of the structure and, more faintly, the ethical complexion of moral commodification are visible in Thompson's great essay. As for structure, *a thing or a practice gets morally commodified when it is detached from its context of engagement with a time, a place, and a community* and it becomes a free-floating object.[43] Literally, of course nothing within our grasp can exist without some spatial, temporal, and social coordinates and causal lines. What happens in moral commodification is that the context of a thing or practice becomes "less transparent" and less "direct," as Thompson puts it. In the case of the eighteenth-century English wheat, "a more complex network of intermediaries" displaces the place, the time, and the people that once were within the understanding and interaction of all involved.

Moral vs. Economic Commodification

This kind of commodification, *moral* commodification, took on planetary dimensions through the Industrial Revolution. The practices and products of the household, of the artisans, and increasingly of the farmers were taken over by mechanization. *Mechanization Takes Command* as Siegfried Giedion famously put it.[44] The mechanization of agriculture and bread making is one of Giedion's topics.[45] More than half a century ago, when Giedion wrote his book, vast, complex, and often distant machineries had come between bread and people's understanding of the where, the when, the how, and the who of its production. In the sixty years since the writing of *Mechanization,* machineries have significantly grown in complexity, immensity, and therefore in opacity to ordinary intelligence.

Typically, then, every commodity rests on a machinery, and moral commodification is always twinned with mechanization. Marx, in embracing the Industrial Revolution as such, endorsed moral commodification though he did so with evident misgivings. He was well aware that through moral commodification "all that is holy is profaned."[46] But the social relations, displaced by mechanization and commodification, were of course unacceptable to Marx. He could see no alternative to moral commodification, and so he embraced it, much as he rejected economic commodification. There are contemporary illustrations as well of the distinctiveness of moral vis-à-vis economic commodification. Farmers' markets are always economically commodified, but not always morally so. In Missoula's market, the local, temporal, and communal bonds are evident and vigorous. You can only sell what was locally grown, roughly within a few hours of Missoula. The market reflects the progress of the seasons with seedlings, rhubarb, and lettuce first, then the peas and peppers, finally the tomatoes and corn. Social interaction is central. The market is in fact the place "where one-hundred-and-one social and personal transactions" go on, where news is passed, rumor flies around, and politics is discussed.

There are also examples of moral commodification without economic commodification. For certain kinds of tourists, distant and spectacular nature and culture are just so much raw material for endlessly taking pictures, films, or videos. Once back home, the results are edited and enhanced. From then on the vacation experience is available regardless of the place, the time, and the people that constituted the setting and origin of what has in fact become a commodity without a market.

A more troubling example along the same general lines is self-commodification. There is of course economic or commercial self-commodification where a woman sells her eggs (legal in the United States) or a man sells one of his kidneys (illegal in the United States). The particulars of such transactions will persuade you that human eggs and organs should both be in the gift economy.[47] At any rate, strictly moral self-commodification is cosmetic, the kind that people undertake and undergo through cosmetic surgery, Botox injections, and the like. An analogous transformation of one's personality is accomplished through designer drugs. This sort of moral commodification is primarily temporal, an extinction of one's history, of the record one's life has inscribed on one's face and body. If you value that history, you will refuse a face-lift as Robert Redford has done because the "trade-off is that something of your soul in your face goes away," as he put it. "You end up looking body-snatched in the last analysis. That's just my view. It's not necessarily a popular view."[48]

Although economic and moral commodification are always conceptually and can be factually distinct, there is an elective affinity between them and hence a large overlap. Trade, as Thompson's essay has shown, can take things far from their original spatial, temporal, and social context. Conversely, once the ties of the engagement that a thing or a practice has with a place, a time, and a social setting have been severed, the thing or practice gets the mobility that lubricates commerce.

Commodification and Ethics

There are, at any rate, evident losses in moral commodification. Certain aspects of life become "increasingly impersonal" as Thompson noted. But what he failed to see is that this can also be seen as a gain. Personal relations can be burdensome. When you follow the development of personal contact among ordinary people since the beginning of the modern era, you move from personal encounters and visits to letter writing; from there to the telegram, the telephone call, to the answering machine and caller ID, to e-mail, instant messaging, and more. None of the successors has entirely extinguished its predecessor. What has been added by each step, however, is more control over other people and an easing of another person's imposition, growing disburdenment from having to be there, to listen, to stay when all this may be a pain. Moral commodification of personal interaction has been a gradual process that has reached an important further step in those encounters that are entirely mediated by information technology. The "other" you meet on the Internet leaves you free of burdens and obligations. Exit from an acquaintance is costless. The process will reach its completion in perfect virtual reality where you can have a person that is perfectly detailed according to your preferences and entirely at your disposal.

Surrogacy, even if we were to think of it as a blameless and socially just enterprise of economic commodification, is an ethically troubling practice from the standpoint of moral commodification. If it is done simply to spare the woman of the contracting party the trouble of bearing a child, there is the loss of the most intimate relation a person can have to the emergence of another person, a loss disguised as liberation and convenience. If the woman is unable to bear a child, keeping surrogacy out of the market and within the gift economy best preserves the personal bonds of obligation and gratitude that are ever threatened by moral commodification. In the case of surrogacy, then, moral commodification can be kept at bay by prohibiting economic commodification, a policy we have embraced when it comes to donating organs.

Thus a promise of liberation propels moral commodification. The freedom in question is not political but personal in the way just described. It is, moreover, real in the sense that moral commodification liberates us from the claims of things. Commodified food frees us from the burdens of harvesting, cleaning, preparing, and cooking. Most important, there is, as some of the examples above suggest, a promise of prosperity along with that of liberty. The Internet is often praised for the richness of personal encounters and experiences it provides.[49] Commodified food is available in much greater variety than what you could hope to cook up at home. As for quantity of food, it has slowly turned from a blessing to a curse.

Apart from variety, what moral commodification promises is the prosperity of superbly pure and unencumbered pleasures. Although we all praise, in the abstract, personal contacts through visits, the thought of having to get up, drive across town, and find our parents in a possibly morose mood makes us turn to a phone call to see how they are doing or, better yet, send them a cheerful e-mail message. Similarly we generally applaud the practice of preparing a meal from scratch and of serving it at the dinner table, but when actually confronted with the task, it looks forbiddingly cumbersome and tedious. Better to order in gourmet food and let everyone eat what, when, and where he or she likes.

Commodity in the sense of pure pleasure recalls the older sense of the word.[50] A commodity is something that is commodious, that is, convenient and comfortable. And this feature of commodities also highlights the specific character of contemporary consumption. To consume in this sense is to take up and enjoy the pure pleasures of moral commodification. What has to be kept in mind throughout is the fact that it is the character of commodities that qualifies the liberty and prosperity they promise—it is their free-floating availability.

But what is specifically moral or ethical about the kind of commodification I have been discussing? The promise of liberty and prosperity that propels commodification is extended as a promise of the good life. In part this has been fulfilled in ways that every reasonable person will applaud. To be freed from hunger, disease, and cold, to have access to education and culture—these are undoubtedly good things. It is also true that much of the planet's population is still awaiting the blessings of basic liberty and prosperity.

Is moral commodification, when used constructively, blameless after all? The blameless-blameworthy distinction, if crucial elsewhere, is too crude here. Although this will invite charges of Luddism, dystopianism, romanticism, and worse, it needs to be said that there are *always* moral losses in any kind of moral commodification. But in constructive applications, the moral gains *unquestionably* outweigh the losses.

Consider infectious childhood diseases. For those of us who went through them they were memorable occasions. Parental concern was never more intense. Fever cast us into fantastic regions. The everyday world never looked as charming as from the distance of the sickbed. There was depth and resonance to these experiences. And yet it would be morally outrageous to want those diseases back. Martha Nussbaum has rightly taken exception to the anthropologist who took the British to task because, when they introduced smallpox vaccination into India, they undermined the cult of the goddess who used to be invoked for a cure.[51] When, to the contrary, we look at the kinds of settings that are paradigmatic for contemporary culture—the middle class of the advanced industrial societies—it is questionable whether recent advances in moral commodification have in fact been conducive to the good life.

There is a normative and an empirical dimension to this question. The moral conception of commodification allows us to answer it with a kind of conceptual structure that takes us beyond the hand waving we find when Luddites and technophiles go at one another. We can bring the normative dimension into relief by contrasting the moral conception of commodification with the standard positions in contemporary ethical theory. Rights and liberties theorists (so-called deontologists) hold that the state should guarantee basic rights and opportunities, but must be neutral when it comes to conceptions of the good life because it is impossible to determine and it would be oppressive to enforce the "good life."[52] Utilitarians (one breed of teleologists) disagree and claim that we all know what the good life is. It's a life of pleasure, and the state should support the maximizing of aggregate pleasure.[53] Virtue ethicists (the other major kind of teleologists) agree with utilitarians that we know the good life. It is however, the life of virtue and moral excellence, not of pleasure, they say.[54]

What is the response from the standpoint of moral commodification? Contrary to the rights and liberties theorists, the state does not and cannot leave the question of the good life open, and states around the globe do in fact promote, and vigorously so, the version of the good life that is tied to economic and moral commodification. But is this in fact the good life? Contrary to the utilitarians, the life of pleasure can be injurious, and the life dedicated to the production and consumption of moral commodities is in fact debilitating. In disburdening us from the claims of people and things it leaves us isolated, passive, and enervated. Virtue ethicists advocate an emphatically different version of the good life, but in their preoccupation with rival theories they have neglected to consider carefully the moral setting and its attractiveness within and against which they preach the acquisition and exercise of virtue. Hence an abstract and elevated air has been clinging to virtue ethics.

The Dubious Pleasures of Commodification

Once moral commodification has alleviated misery and provided for the fundamentals of life, and when it begins to sweep everything before it and to colonize the centers of our lives, it becomes ethically debilitating and objectionable. A life, typically divided between labor, reduced to a mere mans, and leisure, devoted to the consumption of commodities, is not worth living. It is, to be sure, not the worst kind of life. It is superficially pleasant and commonly lived by decent people. It does not deserve moral outrage. But it must provoke sorrow at the widespread loss of vigor and engagement. What *should* provoke outrage are not the intrinsic quality of the commodious life but the consequences of the self-inflicted numbness of that sort of life. Especially in the United States, the pleasures of consumption have made people blind and deaf, as we have seen, to the global misery they could alleviate without severe deprivation to themselves.

Empirical evidence now supports and meshes well with the normative case just outlined. More indirect support comes from the massive evidence Robert Putnam has gathered to document the decline of civic-mindedness in the United States.[55] The increasing density and sophistication of the machineries of administration, commerce, and communication have made selfless devotion to the common good less necessary, I have suggested before, while at the same time the pleasures of consumption have enveloped people in a cocoon of comfort and indifference.

Strikingly direct evidence comes from research in hedonic or positive psychology I have considered when discussing the measurement of happiness. What the notion of moral commodification adds to that account is an explanation of why we today are in a uniquely precarious situation when we pursue happiness. The key notion in this explanation is pleasure. The word has had a wide and varied history as Lionel Trilling has pointed out.[56] In contemporary psychology, pleasure has a specific sense. Pleasures, Martin Seligman has said, "are immediate, come through the senses, and are momentary. They need little or no interpretation. The sense organs, for evolutionary reasons, are hooked quite directly to positive emotion; touching, tasting, smelling, moving the body, seeing and hearing can directly evoke pleasure."[57]

This is by no means an uncommon meaning of pleasure. Throughout history and across cultures, there has been wide agreement that the status of pleasure is unproblematic when we think of pleasure as the moment of delight. In this sense pleasure is immediate and self-warranting. No one needs instruction or persuasion that pleasures are attractive and fulfilling at the instant we experience them. It has always been the context of pleasures that has caused misgivings and, on occasion, has driven hedonists to despair and pessimism.[58]

Contemporary pleasures are situated in a context that is distinctively structured by commodified availability. Economic commodification makes pleasures available in the sense that they are widely affordable in the technological societies. Moral commodification makes pleasures available in the sense that they obtain a singular ease and purity. But all pleasures are haunted by the curse of adaptation. The more we have of a pleasure, the more its power declines. We get used to pleasures; they delight us for ever-shorter periods. But this is a lesson untutored individuals and a thoughtless culture are unable to learn. The natural psychological disposition is to overestimate the utility of new pleasures and to mistake the real but temporary thrill of a sweeter soda pop, a larger portion of fries, a faster car, or a larger television screen for an enduring gain. When once again we are left disappointed, we search for the next and stronger thrill—we are on the celebrated hedonic treadmill.

The availability of commodities aggravates this "inviolable neurological fact of life."[59] Because they are economically commodified, our pleasures lack the spacing, timing, and constraining that in earlier times preserved the savor of pleasures. Because they are morally commodified, they are uniquely disengaged and lacking in depth and resonance. All this aggravates our obsession and disappointment with pleasures. It leads to an epidemic of depression and a decline of happiness.[60]

What should we do? Lead a life devoid of pleasure? Draw another line? This time between moral commodification and what? The answer is not to find a line, but to remember and invigorate those centers in our lives that engage our place, our time, and the people around us. In the personal sphere these are focal things and practices such as the culture of the table. In the public sphere they are centers of communal celebration such as farmers' markets.

Here too the empirical evidence is illuminating and in fact encouraging. Enduring joy is intrinsic to the engagements of focal practices and communal celebrations.[61] Moreover, pleasures embedded in engagements will not betray us. And finally, if we in the affluent countries lead lives that are good as well as pleasant, we can get off the hedonic treadmill and use our resources to be a global force of genuine liberty and prosperity.

CHAPTER THIRTEEN

The Economy of the Household

The Grounding Virtues

Commodification can be subversive. It can take things and practices that are dear to us, reduce them to slick and available merchandise, and sell them back to us. But when such things or practices return to us as commodities, they have lost their centering and consoling power. Preparing a meal and gathering around it at the dinner table may have been an anchor and focus in our troubled world. A fine meal? Why not order it in from the Gourmet Shop that has just opened up down the street? And why not order for each of us their favorite dish? And since Pat is not yet back and Leslie is not quite ready, why not let everyone eat at his or her leisure?

The subversive power of commodification is so strong because it is concealed, and it is concealed because it moves along a smooth gradient from good to bad and from harmless to injurious. There is no red line on that gradient that warns us: So far, but no farther. To transform health from something that is painfully and perilously wrested from whooping cough, diphtheria, or scarlet fever into the commodious effect of vaccination is undoubtedly a good thing. To convert the water supply from a social event at the village fountain, a laborious transport to the home, and an occasional source of cholera into a safely and easily available commodity is, on balance, a good thing as well. A gourmet meal ordered in when both parents return a little late and more than a little harried is not a bad thing either. Reading *Charlotte's Web* to your children is a good thing. So why not let them watch the video afterward? And then a video of "an idyllic northern lake full of wildlife," the "Einstein" videos—and before you know it you find yourself on the slippery slope to "SpongeBob" that Lynn Neary has described. So where to draw the line?

Commodification does not simply move of its own along this gradient. It is being pushed by business, and those who do recognize the subversive side of commodification often assume its power is entirely due to business pressure. Hence they conclude that, unless capitalism is replaced or reformed from the ground up, there is no hope of reining in commodification.

The kernel of truth in this is the insight that commodification is not an independent and overpowering force that has its way with us; rather commodification is something that people do, or not do, or are able to modify. It is something we have fundamentally and implicitly agreed on. We, or nearly all of us, in this country, are implicated in it. How such profound and influential agreements well up from the depths of history is impossible to say. In retrospect we can see favorable conditions and evident triggers of this development, but there is nothing like a specifiable set of sufficient causes, nor is there a historical pattern that tells us what exactly came before commodification and what will follow it. The moral challenge, in any case, is not causal analysis but paradigmatic explanation and, once commodification is explicit and recognized, taking responsibility for our implication in it and removing it from the center of our lives.[1]

This in turn requires a modification of capitalism, nothing more or less. The free market, the heart of capitalism, most vigorously engages the efforts of people and most efficiently allocates resources and distributes goods and services. But the freedom of a healthy market has always been hedged by political institutions, by health and safety regulations, good faith and credit requirements, competitive rules, and the like. These political and social background institutions are well understood if not widely agreed on between conservatives and liberals. What is missing is an appreciation of the material and cultural, the real, background conditions, that are governed by Churchill's principle. The task is to turn that principle in favor of the good life and the good society. Unless we get to be schooled and practiced in realizing Churchill's principle on behalf of moral excellence, this won't happen. In other words, we have to acquire the virtues of economy and design.

There are three things practitioners of those virtues have to master. The first is a lucid understanding of the moral meaning of commodification. Without it, well-intentioned efforts get confused at best and subverted most of the time, pulled by the gravitational force of commodification. The second component of economy and design is the obverse of the first. Since moral commodification severs things from their engaging spatial, temporal, and social settings, reforming our tangible environments comes to restoring the ground of things and practices to a depth where they can again engage us in the fullness of our capacities. Finally, practitioners of economy and design need a certain kind of literacy, acquaintance with landmarks of homes and public places that are conducive to moral excellence.

But what's so great about being grounded and about the grounding virtues? After all, being grounded can mean to have your wings clipped. Confinement, being stuck in the mud, was one of the conditions modern technology released us from. Aren't mobility and variety, the ability to experience different places and cultures and to enjoy different phases and roles in one's life, hallmarks of an exciting and successful life? Is it not judgmental and puritanical to want to suppress this sort of freedom and richness? It is certainly true that the life of unencumbered richness and diversity rules the imagination of our society and culture.

Several things need to be remembered in reply. The first is about moral discourse. At its most important it is always a plea, and especially in a democracy it must be a plea; it cannot be trickery or imposition. More particularly it is a plea on behalf of a certain kind of life. The challenge for the philosopher is to be an effective advocate. But at the end of the day, when all the presenting and persuading has been done, that way of life has to recommend itself in its own right.

Next we have to remember that when it comes to the quality of life, we, whether as individuals or as citizens, cannot be neutral. Our lives always have a certain character, and we are responsible for it. As it happens, contemporary American life is heavily tilted toward commodification and consumption. Hence wanting to level the field of life and countering commodification is not on its face unreasonable.

Then we also need to recall what the great moral traditions from the Greeks to the Blackfeet, what recent social science, and what the benevolent parental standpoint are telling us with one voice:

It's a life of virtue and engagement, not a life of commodious pleasures, that answers our deepest aspirations and makes us profoundly happy. Pleasures are fine, even crucial. But they need to be integrated again into a world of focal things and practices and of communal celebrations.

And one final point. The currents of language and the rise to prominence of certain words sometimes foretell a shift of culture. To say of a person that he or she is *grounded* has lately become a way of commending that person. The suggestion is that such persons are steadfast, sure of their place in the world, and unassuming in their convictions. They are people we look to for counsel and consolation.

Economy and design are the virtues that help us to ground our lives. Such grounding is the least understood task of ethics we face. Therefore the grounding virtues have a special moral status. At the same time, the good life is a coherent whole, or it is not good. Wisdom, courage, friendship, and grace are needed if the ground of life is not to remain fallow.

The Jefferson and Wright Households

Your home grounds your life most immediately, and it's also the sphere of life where you have most discretion and responsibility. There is nothing you and your spouse can do by yourselves about the Interstate highway that splits your neighborhood. But you can start rearranging your home this evening. Most of us have to do this within limits, especially if we are renters or of moderate means.

Thomas Jefferson and Frank Lloyd Wright were architects and so had exceptional freedom in how to shape their houses. They kept reshaping them as long as they lived in them, Jefferson over his entire adult life, Wright, in his first house, for two decades. "[A]rchitecture is my delight," Jefferson is reported to have said, "and putting up, and pulling down, one of my favorite amusements."[2] One of his associates, John H. Howe, said of Wright's design process: "He couldn't wait to tear it down."[3] So if economy is the virtue of shaping your home vigorously and well, Jefferson and Wright were exemplary practitioners of it.

Although neither Jefferson nor Wright was ever without financial worries, both commanded enough prosperity to build large and elaborate houses for themselves and their families.[4] At length, both houses came to something like a completion, and for our purposes it is most instructive to look at them at that stage. In their final form, both houses are a foil and a model for contemporary economy—a foil in showing features we cannot adopt, a model in calling our attention to what a home should be today.

One thing that immediately troubles us about Monticello, Jefferson's house, is the location of the kitchen, icehouse, brewery, smokehouse, stables, and of the quarters for the slaves who operated these "dependencies." With misplaced ingenuity, Jefferson had them dug into the sides of the hill that the mansion stood on. They were connected to the long axis of the house by an underground hallway. The entire U-shaped structure was invisible from the mansion. Terraces concealed the roofs of the two L-shaped parts on either side of the house.

The Wright house too has a trace of such servitude. Between the kitchen and the dining room is a small chamber for the maid.[5] Today technology allows us to be more comfortable and more egalitarian than Jefferson and even Wright were. We have technological devices to do our own cooking, washing, cleaning, refrigerating, provisioning, and traveling.

Equally important, today's greater egalitarianism makes patriarchy as unacceptable as servitude. Jefferson was a patriarch by default. After "ten years of unchequered happiness" Martha Wayles Jefferson died, and there was no longer a matriarch with whom Jefferson could have shared his say-so.[6] In his final years at Monticello, after his presidency, Jefferson was distinguished by venerable age as much as by being a man. Wright lived through the waning of patriarchy. In his early marriage he thought of himself unself-consciously as head of the family: "The young husband found that he had his work cut out for himself. The young wife found hers cut out for her. Architecture was my profession. Motherhood had become hers. Fair enough, but it was a division."[7] In the 1930s, Wright began to recognize the aspirations of a woman and her role in the house. "Since she wanted a compact, efficient home without servants," Robert Twombley tells us, "Wright moved the kitchen from the back of the house to the center, calling it the 'work-space,' from which she could run the entire operation."[8]

Economy is a virtue that turned out to be more problematic for Wright than for Jefferson. Wright's Prairie House was a contribution to the shape of private residence that was more celebrated than the structure of his house in Oak Park. Unlike the latter, the Prairie House represented the "open plan" where kitchen and living and dining rooms were one or at least flowed into one another without the divisions of walls and doors. The open plan was uncertain from the point of view of economy. Wright intended it to encompass and unite the several activities of the family. But the plan could also lead to the loss of order and the dispersion of activities.[9] In the event, television invaded the open area and resolved the uncertainty. The axis of couch to television set overwhelmed and absorbed all else. This problem is still with us and is being once again aggravated by the next stage of media technology—greater availability of content and more captivating forms of presentation.

A closer look at the houses and lives of the Jefferson and Wright families reveals that with all the uncertainty of what to do with families and with all the stupor that today can flow from digital screens, there are some monuments of economy and virtue left. Nothing about them is exclusive or mandatory. We should certainly be open to experiment and innovation and on the lookout for new things and practices. But neither should we toss out what has served us well and serves so to this day, nor should we indulge novelty uncritically.

Economy and Friendship

Economy undergirds virtue with reality. Friendship today badly needs things to center and sustain it, friendship in the eminent sense of marriage (whether heterosexual or homosexual) and in the wider sense of hospitality. That marriages are not doing well everyone knows. But inviting friends to dinner has suffered too, and so has the family dinner.[10]

Monticello has a formal dining room, not particularly large by McMansion standards, though it could be extended into the adjoining tearoom. Both breakfast and dinner were signaled by the bell and taken communally with Jefferson presiding. The Wrights too took pleasure in dining, and Wright had shaped a magnificent space to dine in. It is both warm and solemn with a red-tiled floor, golden-toned paneling, a heavy table in the middle surrounded by the chairs whose straight and high backs define a space within a space. At the short sides of the dining room are a half octagon on one side and a fireplace at the other, exactly the shape of the Monticello dining room if we include the half-octagonal tearoom.

For Jefferson a fireplace was a practical necessity, and octagonal shapes were convenient at a time when illumination was weak and expensive and dark ninety-degree corners were a nuisance. Wright for his part recognized the symbolic force of fireplace and octagon. As for the fireplace, Neil Levine has captured its significance in the Wright house perfectly: "The round-arched brick mass of the fireplace, set deep within an inglenook, forms an inner sanctum. Removed from the main flow of space, it becomes almost purely symbolic in function, providing a constant reminder and stable image of the mutual affection that ideally guides the family life revolving around it."[11] Octagons, full and half, were emphatically enclosing spaces. There were five such in the Wrights' house (counting the dome over the drafting room) and an equal number in Monticello. Even when Wright had dropped the outer enclosure of the dining space in favor of the open plan, he continued to mark the inner space by means of high-backed chairs and, in some instances, by emphasizing the stature of the table. He placed pillars or standards at the corners of the table and topped them with lights or flowers.[12]

Both Jefferson's and Wright's friendship in the wider sense of hospitality equaled their economy. They filled the spaces they had created with friends and guests. Jefferson's daughter Martha had to take over as hostess at Monticello and did so with admirable devotion and resourcefulness. When she remarked, however, how little time this left her with her father, to whom she was even more devoted, Jefferson replied that the obligations "are laws we cannot repeal."[13] Wright's son, John Lloyd Wright, remembers this of his father's hospitality: "Papa's parties were best of all. He had clambakes, tea parties in his studio, cotillions in the large drafting room; gay affairs about the blazing logs that snapped and crackled in the big fireplace. From week to week, month to month, our home was a round of parties. There were parties somewhere all of the time and everywhere some of the time. Bowls of apples and nuts, great jars of wild flowers were everywhere."[14]

The space of hospitality extended beyond the compass of the dining room. In Monticello, guests would enter on the west side a two-story entrance hall that told them of the landmarks in Jefferson's life, busts of Voltaire and Hamilton, a painting of the signing of the Declaration of Independence, a Mandan buffalo robe and other Native American artifacts Lewis and Clark had brought back from their expedition. From there, the guests might move along the main axis of the house to the other great room of the house, the parlor. It was filled with art and musical instruments. It was the place for play, celebration, and reading.

The living room in the Wrights' house is warm in color and open in its layout. In one corner it extends into a half octagon of windows, enclosing the space inside and disclosing the yard outside with the library across the yard. For large occasions, the Wrights used the drafting room, a square, two stories high, capped by an octagonal dome that is artfully contained by a system of chains; they similarly used the magnificent playroom on the second floor, overarched by a barrel vault, anchored by a balcony at one end and a fireplace at the other, opening on bay windows on both sides.

Economy and Wisdom

The thing that helps us realize the virtue of wisdom is the book. This may seem like a quaint claim in the age of electronic information. The Internet is in fact a marvelous source of information. But it generates knowledge poorly, and it is clueless as a guide to wisdom. The mass of information that a computer puts at our fingertips presupposes rather than produces knowledge. You have to know what information you need and where you are likely to find it before a search engine becomes useful. Then, to be sure, information can refresh and deepen knowledge. Knowledge is information appropriated and integrated. But the web of knowledge can still be, and usually is today, without patterns of orientation and centers of meaning. Wisdom, to the contrary, is the comprehension of the crucial features of the world and of our place in the world. There is no consensus on that, but there are creditable pieces and proposals that deserve our careful attention.

It takes a book to present a worldview. A book has the right length and the right genesis. Some books today are written overnight and little more than long pamphlets, and they surely have their place in the national conversation, but typically a book is the labor of some years. The prospect of writing a book lends an author the pensive mood and the willingness to labor that a contribution to wisdom requires. At the receiving end, readers have before them a communication that is austere, well bounded, and silent. Page after printed page looks physically spare and severe. It takes mental engagement to bring print to life and to make it speak. But it can come alive, drawing its vigor from the readers' experiences and, if things go well, ordering and deepening the way readers understand their world.

Books as concrete things, pages bound together and enclosed by covers, differ from the information on a computer where information is inevitably in closest proximity to more and other information. The slightest of human actions, a mere twitch of a finger, takes us from a classic text to sports, the market, and entertainment as it did in the case of Lynn Neary and Mark Patinkin—the quest for a child's instruction and the quest to write a novel were waylaid by distracting entertainment.

A book instructs us quietly. It does not preclude or intrude on conversations the way television or video games do. To be sure, reading is favored by silence, but it is a silence that can be broken graciously as when we read a passage to a loved one or ask for an explanation of a phrase. Books, moreover, do not disappear into nothing or into disks that are humanly inaccessible. Books can gather into a library that marks the stages of our life and constitutes the visible and abiding background of knowledge and wisdom.

Jefferson professed to John Adams: "I cannot live without books," and he could still hope to assemble books into a library that would reflect all essentials of human knowledge in an orderly arrangement.[15] He organized his holdings under three headings, adapted from Francis Bacon's system: Memory, Philosophy, Fine Arts. These divisions faintly recall the medieval triad of the true, the good, and the beautiful. More particularly, if still approximately, the three categories comprised factual knowledge, knowledge of values and theories, and knowledge of the arts. Each of these categories was subdivided into roughly fifteen "chapters."[16]

Wright had to be more modest and selective. His passions were literature and art. He built an octagonal library for his books as part of his studio, but it became a place for receiving clients rather than reading books. Jefferson's library in Monticello is a half octagon opening on a square. Reading, however, was done in the great parlor and was often communal. One of Jefferson's granddaughters remembered that, "[w]hen the candles were brought, all was quiet immediately, for he took up his book to read; and we would not speak out of a whisper, lest we should disturb him, and generally followed his example and took a book; and I have seen him raise his eyes from his own book, and look round on the little circle of readers and smile."[17]

Economy and Grace

Wisdom naturally shades over into grace, the virtue of the arts and, for many of us, of religion. Jefferson's Fine Arts categories contained these chapters or subdivisions:

33. Poetry. Epic
34. Romance. Tales—Fables
35. Pastorals, Odes, Elegies
36. Didactic
37. Tragedy
38. Comedy

Poetry, fiction, and drama can be enjoyed through silent or voiced reading. But perhaps grace comes most fully into its own in music. In our culture, music is widely present. You won't find a construction crew without a boom box that has a country and western or golden oldies station blasting, nor will you find many teenagers who have failed to spend a lot of time in front of MTV. Car radios and iPods envelop people with music. But such musical presence is like a colorful sonic curtain. No one knows what goes on behind the audible surface. Nothing musical happens in front of the curtain.

For musical grace to come to life, the curtain needs to be raised. We have to take up instruments, learn how to play, and allow our engagement with the instruments and one another to fill our lives with the rhythms and harmonies of music. Jefferson was a competent violinist till he broke his right hand in a riding accident in France. A music stand and a harpsichord are found in the parlor of Monticello to this day. Jefferson's wife played keyboard instruments, and his daughter Martha became an accomplished pianist and harpsichordist too. Jefferson called music the "favorite passion of my soul," and it "furnishes," he said, "a delightful recreation for the hours of respite from the cares of the day, and lasts us through life."[18] From his catalogue and the sheet music that survived we know that the Jeffersons played a wide variety of instrumental and vocal music from the baroque and early classical period.[19]

Wright's wife, Catherine, was a fine pianist as well, and the Wrights saw to it that each of their children would learn to play an instrument or to sing. The children played together as a little orchestra, Wright tells us.[20] He played the piano himself. It still stands in a niche under the balcony of the playroom. John Lloyd recalls how vital the grace of music was for his father: "I remember the way he improvised chords and cadenzas. In quiet moods he would improvise soft melodies full of tenderness, soothing, gentle, like quiet waters. Suddenly in complete abandon of all responsibility he would take bold flights in rhythmic vehemence. At such times crescendos of pent-up power poured fourth, then a relapse into diminuendos strangely elusive as himself."[21]

Economy Today

What are we to take from Jefferson's and Wright's economy and the way it supported the virtues of their families, the virtues of friendship, wisdom, and grace? Friendship is most within reach, wisdom less so, and grace is still more remote from our present habits. Dinner is clearly the most hopeful focal point of a marriage and a family. Hunger reminds us every evening that there is an occasion for breaking bread together. The tenacity families have shown in clinging to a dining room in their house suggests a readiness to make dining a virtue, a skilled moral practice. Those of us who lack a formal dining room can follow Wright's example in his Prairie Houses and highlight and distinguish the dining area. What should make us still more determined to give the culture of the table a central place in our homes is the support social science has given our intuitions that dining together is good for body and soul, particularly so for those in the most vulnerable age—our teenagers.

Just as we can make the setting of the dinner table more prominent and inviting, so for the culture of the word, for the reading and conversing that allow us to center and appropriate our world more fully. Here the chief task is as simple as it is unimaginable for many—move television out of the living room or the family room to another room so that our default behavior is to sit down to talk and to read rather than to munch and to watch.

Grace is surely the most difficult virtue to plant and cultivate. It's important to realize right at the start that the obstacles are not economic. Three-quarters, at least, of American families have the money to acquire an instrument, say a recorder or a guitar. And they have time to learn, to practice, and to play regularly if they watch less television. Instruction, finally, is available from schools or from methods that allow you to teach yourself.

Grace, of course, does not require that everyone is a musician, only that living music is a presence. Nor does grace require music exclusively. Poetry, dance, painting, weaving, ceramics, and more can generally be as graceful as music. Economy must play its role here. Instruments must have a ready and prominent place in our primary room so they can invite and remind us to take them up regularly. Studios for artwork must be close and comfortable.

What grace needs most, though friendship and wisdom require it too, is the virtue of courage, not courage as reckless abandon or the fearless facing of dangers, rather more in the sense of steadfast persistence against comfort and convenience—the virtue that may better be called fortitude. The good life is within reach for most of us. What comes between us and the good life are the unencumbered and seemingly sweet pleasures of technological affluence. They are advertised to us. They need no lengthy introduction or training. But at length they leave us empty. Virtues seem forbidding in comparison, or worse, they have become unfamiliar and invisible. They need introduction and advocacy, and once encountered, they need fortitude to be acquired and sustained. Above all they need a favorable setting, one that reminds us and invites us to be faithful to them. Then they will reward us deeply and enduringly.

Concluding this chapter, we can see this conjunction of habitat and virtue in Jefferson's household. He shaped his house and afterward his house shaped the course of his daily life. To be sure, the Jefferson household must serve us as a foil as much as a model for our own economy. Jefferson's economy was a feudal arrangement, based on slavery. It failed to include in its affection the children Jefferson had with Sally Hemings, as Brent Staples reminds us.[22] Moreover, it erected an ostentatious building in the splendid isolation of a hilltop in the Virginia countryside. It embodied Jefferson's agrarian ideal, a life sustained, more or less, by agriculture. We, to the contrary, aspire to equality. At our best, we promote the pleasures of density and urbanity. And we recognize the necessity of industry and commerce as Jefferson grudgingly came to accept it too.[23]

What makes Jefferson's household a model for us is the restful sort of excellence it represents. Resting our life with calmness and the confidence that we have done our best is something we are badly lacking in contemporary American culture, where excellence is typically restless and perpetually haunted by the next project and by the anxiety that next time we will fail. Just look at the country's elite in politics, business, the arts, and in the sciences.

The lapidary representation of secure and enduring excellence was Monticello itself. In its symmetry and harmony it is a classic building. The guiding spirit of Jefferson's architecture was the Renaissance architect Andrea Palladio (1508–1580). But just as the civic republicanism of the Renaissance finally reaches down to the Roman republic and Athenian democracy, so Jefferson saw Palladian architecture as an embodiment of Athenian virtue and Roman rectitude. In a letter to the housejoiner James Oldham, Jefferson contrasted the "chaste architecture" of the Palladian style with "the barbarous and tawdry fancies of each individual workman."[24]

Monticello was both grounded and innovative. It represented both "memory and invention" as all good architecture must according to Robert Campbell.[25] Most of it was built from local materials by local labor (much of it by Jefferson's slaves). It grew out of the "little mountain," and the household it contained was nourished by the fruits of surrounding agriculture. As important, it was filled with pleasure that rested on virtue. Jefferson consciously distanced himself from the turbulences of daily politics—so circumscribed and comprehensible by our lights. "I have given up newspapers," he wrote to John Adams, "in exchange for Tacitus and Thucydides, for Newton and Euclid, and I find myself much the happier."[26]

Jefferson and Adams had been bitter enemies during Adams's presidency. Thus the correspondence with Adams was itself a testament to Jefferson's peaceful engagement with his world. This pacific embrace did not, however, include Jefferson's children with Sally Hemings. He gave them their freedom, but not his affection.[27] The children of his daughter Martha, to the contrary, had his devotion. He taught them, he encouraged them, he led them in games, he showered them with the gifts that encourage virtue, and he was affectionately remembered for it all. Thus a great building was filled with life and some grace.

The Wright household too is both foil and model for our economy. In 1898, Wright built his studio next to his residence and moved his work from downtown Chicago to Oak Park. Thus he restored the proximity of work and life that used to give families and especially children a fuller and healthier sense of how the world and the expertise of parents establish rightful authority.[28] But this ideal is not within the reach of most of us, and it did not work well in the Wright household. The children came to be a nuisance in the studio area, and Wright, even on his own record, could be a capricious if not also cruel father. The children in their turn were apparently anything but poised and polite.[29] Whatever beneficial shaping the building did to Wright after he had shaped his house, an enduring conjugal shape was not among the benefits.

Wright was exemplary, however, in taking responsibility for the virtue of economy. He placed his house and studio into the larger urban fabric.[30] He questioned, rethought, and reformed the nature of the family residence. He recognized and celebrated the place where it was built, the materials it was made of, the octagons that enclosed it, the fireplaces and dinner tables that constituted its focal points. If he was more successful in shaping his buildings than in allowing them to shape him, the force of Churchill's principle was not entirely lost on the Wright household as Wright's son John Lloyd testified: "It was Dad's desire that his children should grow up with a recognition of what is good in the art of the house. He believed that an instinct for the beautiful would be firmly established by a room whose simple beauty and strength are daily factors. And Dad was right."[31]

CHAPTER FOURTEEN

The Design of Public Space

Space and Public Policy

There is no doubt that we shape our homes and that our homes bank the moral course of our personal lives. But do we as a society take an active role in shaping public space? And if we do, does the shape of public space give moral direction to our conduct? In short, while there is a need for the personal virtue of economy, is there a place for the political virtue of design? Perhaps an answer can be had by first looking at the historical setting of Churchill's principle, then at the imprint of the principle on the boonies of the northern Rockies, and finally at the impact the principle has recently had on the most metropolitan of cities, New York City, and on the focal point of Ground Zero.

"On the night of 10th May, 1941," Churchill said two years later, in an address to Parliament, "with one of the last bombs of the last serious raid, our House of Commons was destroyed by the violence of the enemy, and we have now to consider," Churchill continued, "whether we should build it up again, and how, and when. We shape our buildings, and afterwards our buildings shape us."[1]

The original House of Commons of the sixteenth century had been a chapel dedicated to St. Stephen. It had ever since been oblong in shape and for some time too small in size. Logically, a semicircular and more spacious hall seemed advisable as a replacement. "Logic," Churchill said to the contrary, "is a poor guide compared with custom."[2] The oblong shape with the parties facing one another on benches, Churchill argued, was crucial to the vigor of the two-party system. Limited space would keep nine-tenths of all debates from being conducted "in the depressing atmosphere of an almost empty or half-empty chamber," while "there should be on great occasions a sense of crowd and urgency."[3] Churchill concluded his remarks on the relation of House and conduct as follows: "We have learned—with these so recently confirmed facts around us and before us—not to alter improvidently the physical structures which have enabled so remarkable an organism to carry on its work of banning dictatorships within this Island, and pursuing and beating into ruins all dictators who have molested us from the outside."[4] Should the shape of the House of Commons be credited with Britain's parliamentary government having defeated Philip II of Spain, Napoleon, and Hitler? And after the rebuilding of the House of Commons, did the United Kingdom prosper more than the democracies whose parliaments meet in spacious semicircular halls?

An affirmative argument would be hard to make. The direction of human conduct is the resultant of at least two vectors, of resolve and of "physical structures." The strength of one can make up for the weakness of the other. Steely resolve to do the right thing can compensate for, or even overcome, a weakly supporting or antagonistic environment. But most of us lack that iron will, and we would "improvidently" ignore the force of "physical structures."

The House of Commons in its characteristic shape puts a soft though important constraint on the conduct of politics. Not so the Interstate Highway System on the character of contemporary life. Since 1970, when we moved to Missoula, Montana, we have been taking Interstate 90 to Spokane and Seattle in Washington. In the following decades, the stretch between Missoula and Spokane coalesced into a continuous four-lane highway from pieces that once were precariously connected by two narrow, winding, and sometimes shoulderless lanes. Many a new bridge had to be thrown across the gorge of the Clark Fork River to get the Interstate smoothly to the foot of the Bitterroot Range. Masses of rock had to be blasted away and endless retaining walls erected to take the highway across Lookout Pass into Idaho. The former rail line, now a bicycle trail, that had preceded the Interstate by almost a century, in comparison seems dainty with its narrow roadbed chiseled into the rock and its spidery trestles crossing the rivers.

When we first would travel on a newly installed piece of Interstate, ease and grace seemed to descend on our traveling. We had become masters of the craggy and precipitous terrain. The skill and cunning of engineers and the violent force of dynamite and bulldozers had accrued to us as the power to move with fleetness and assurance. More broadly considered, we were witnessing the completion of an enterprise far more gigantic than the piecing together of two hundred difficult interstate miles. The Interstate Highway System had been conceived and planned, and reconceived and restarted since the late 1930s and was finally put on a solid footing through the Federal-Aid Highway Act of 1956.[5] The history of the half-century since the first beginnings of the System is a chronicle of how tenacious, vigorous, and radical public policy can be in shaping public space. Policy makers did not merely think of the Interstate System as a good idea and a fine thing if they could have it. They were visionaries and zealots on behalf of a sacred cause that had to be nurtured and sustained against inertia, divisiveness, and diffidence. The nearly 47,000 miles of highway that have meanwhile been completed constitute the most extensive and expensive structure that humans have built on earth.

Once the American citizenry had settled its confusions and divisions and made up its legislative mind, the implementation was just about painless and pleasant to most citizens. There were local disruptions where houses needed to be razed, traffic flows diverted, and neighborhoods divided. But at first, any rate, these changes were suffered with good humor; they were a tribute to progress. For most people the new interstates appeared piece by piece as thoroughfares of welcome mobility. Of course every driver had to pay for the construction of the system through gasoline taxes. Yet they were a negligible burden, a token of the enormous affluence that modern technology has supplied to its beneficiaries.

Wallace, Idaho, was the most recalcitrant obstacle to the westward progress of I-90. For twenty years we would look forward to Wallace with dismay and pleasure. Wallace invariably arrested our smooth and pleasant progress, ensnarled us in slow-moving traffic, and more often than not stopped us at Seventh and Bank Street. But inevitably also we were captivated by the courageous grace of the Victorian buildings and by the waning aura of mining and prostitution. At times we stopped to look at the fading splendor of hotels and banks and to immerse ourselves in the talk and smells of a coffee shop, witnesses to the slow and familiar ways of small-town life. What would it be like to grow up in this languishing economy, among these colorful buildings, and amidst these ravaged hillsides? Meanwhile the Interstate was rising on the northern side of town, carved into the mountain here, swooping overhead there, displaying tiled retaining walls, gigantic columns, and elegantly sculpted decks, $42 million worth of effort to construct an overpass of a mile and a half.[6]

On September 7, 1991, we were driving from Spokane back to Missoula. There was no detour from I-90 as we approached Wallace.[7] Before we realized it, we were continuing to cruise along on a familiar Interstate surface, slightly curving left and then right. On that day the heroic substructure had become invisible; Wallace was reduced to a picturesque jumble of roofs and facades. Within less than two minutes Wallace had appeared and disappeared. This experience can stand as a paradigm of the final effect that public policy has had on public space in the second half of the past century. Tenacious and far-reaching public policy has by way of sophisticated engineering and powerful construction left us with a dematerialized and inconsequential experience of our world—public exertion imploding into private ease. The comfort of Interstate travel is the way the hard constraint of highway engineering is experienced in daily life.

The Commodification of Space

The first Interstate exchange west of Missoula connects I-90 with North Reserve Street. In the 1970s and 80s, Reserve Street was a two-lane road traversing mostly open land. In the 1990s, the county commissioners, the federal government, the Montana Department of Transportation, and land developers collaborated to construct the North Reserve corridor. The federal government furnished the money to widen North Reserve to four lanes; the Montana Department of Transportation provided the engineering; the county commissioners saw to the zoning; developers got in touch with box stores and vice versa. It began with Costco, proceeded to the Home Depot, and ended with Wal-Mart, or so it seemed, because then Lowe's followed, and many more. There is a Barnes and Noble, a Target store, a PetSmart, a Staples.

From downtown Missoula, it's a ten-minute drive on I-90 to North Reserve. There is no longer a traditional hardware store in Missoula. The convenience of I-90, Lowe's, and Home Depot draw us inescapably to North Reserve. So do the treasures of Costco, the books and journals of Barnes and Noble, the movies at the ten stadium-seat Carmike theaters, the luxuries of T. J. Maxx, and more. North Reserve draws shoppers all the way from Washington and Idaho.

This amounts to a well-oiled consumption machine, though it is beginning to overheat. In 1980, just north of where Reserve intersects with River Road, 11,540 trips a day were counted. For the reconstructed Reserve Street, a volume approaching 40,000 trips a day was anticipated by 2015. But in 2004 already there were 42,000 vehicles a day. An errand along Reserve that the *Missoulian* writer Ginny Merriam expected to take less than an hour in fact took her two hours and twenty minutes.[8] But even if the traffic machinery of North Reserve worked smoothly, moral problems would remain. "You may have a machine," Churchill said of a more convenient House of Commons, "but the House of Commons is much more than a machine; it has earned and captured and held through long generations the imagination and respect of the British nation."[9]

The machinery of the Interstate Highway System together with the machinery of automobiles has commodified space. Commodification in the moral sense is a matter of more or less, rather than everything or nothing. Time commodified loses its trying and boring extension and at the limit turns into instantaneity. The distance from Missoula to Spokane used to be five hours in a hot, noisy, and bumpy VW bus. Now it is three hours in the air-conditioned and music-filled comfort of a Subaru Outback. In a jet it takes forty-five minutes. A scramjet would take you anywhere in the United States in half that time or less.[10] The compression of time amounts to a compression of space, and, again at the limit, distance yields to be ubiquity. You are instantly anywhere you like; so in effect you are everywhere at once.

Interstates and cars attenuate time and space only by degrees, but those degrees matter or rather dematerialize. They attenuate our interaction with the world. The force of the weather is minimized, the mountains and valleys are smoothed out, the towns recede into scenery, contact with people has evaporated. Commodification by degrees has estranged Missoulians from their city and from one another. We spend more time in our automobile cocoons and the anonymity of large stores, some now with checkout stations without a cashier. Nationally, the transformation of space by interstates is well known. They have eviscerated the inner cities and their vital urbanism; they have distended the suburbs and abetted physical shapelessness and social disconnectedness.[11]

There can be little doubt then that we are vigorously shaping public space through government and business and that the shape of public space channels our moral conduct. The virtue of design comes to taking responsibility for how we shape what we have in common. Had we done so half a century ago, would we or should we have decided not to go ahead with the interstates? Some sort of multiple-lane limited-access highway network is definitely needed in the United States, but we have shown a determined ignorance as regards the consequences of what we have built. We construct a bypass to draw traffic away from a city, but the bypass gathers a new city around itself.[12] We increase the number of lanes and add arteries to make cars move more quickly and smoothly, but the new roads generate more traffic without reducing the old, and soon it takes more time to get from one point to another than before. "It's as if we hadn't learned anything in the last 50 years," David Bernstein, a civil engineering professor at Princeton, has said. "The rule is this: If you build it, they will come. It's called induced demand. Every mile of road you build induces people to drive."[13]

Certainly in the metropolitan areas expressways should have been limited in favor of more and better public transportation. As always, however, the central answer to commodification is not prohibition but realization, the mental version of recognition first, and then the tangible version of building real structures that support celebration. Realization favors celebration just as commodification spawns consumption.

Public Space and Celebration

One such real structure is a baseball park, and again, out in the nowhere of the northern Rockies, Missoula, Montana, seemed on the way to establishing such a place of celebration. A group of inspired supporters surveyed the possible sites and narrowed its choice to an abandoned lumberyard close to downtown on the Clark Fork River that dominates the Missoula valley. The Missoula Redevelopment Agency offered money for the improvement of access. The baseball people joined forces with the ice rink proponents to enlarge the popular base. They tenaciously prevailed on the city council to put a bond issue of $3.5 million for the purchase of the land on the November 5, 1991, ballot.

We had visions of families strolling down the Clark Fork on a summer's evening to the western edge of downtown Missoula, taking their place with some three thousand others on the bleachers in view of the city's gentle towers and graceful buildings, surrounded by the ranges and peaks of the northern Rockies. Missoulians would see games where raw skill and youthful ambition are struggling to acquire the polish of major league finesse. After the game, people would cross the river and walk a few blocks east to reach the heart of downtown, to stop there for a beer or ice cream, perhaps to buy this or that, or just to look and to be seen. For all its virtues, Missoula's summertime and downtown evenings need such vital and focal occasions.

In hearings, letters, and advertisements, the ballpark proponents alluded to this festive vision with commendable force. Of course, they also felt compelled to use the dominant discourse of public machineries and praised the economic benefits of their proposal. The opponents were quick to point out the ballpark's meagerness as an economic instrument and rigorously opposed the very idea of public investment in a final, festive good. The local paper sagely distinguished between needs and wants, declared the ballpark a want, and prudently favored future needs over present wants.[14]

The machinery of municipal government favors inaction when it comes to bond issues. They fail, whatever the outcome, if fewer than 30 percent of the registered voters turn out. They require a 60 percent majority to prevail if fewer than 41 percent of the electorate shows up. The bond issue seemed favored by the fact that Missoula is not a heavily indebted city and that the average property owner would have had to pay only about $20 per annum for twenty years.[15] But in the event, no substantive argument, no electoral regulation, and no financial circumstance appear to have mattered. The bond issue was rejected by a 72 to 28 percent margin.[16] All the bond issues in western Montana failed that day, two for schools, one for a water system, and one for a fire truck, but none so spectacularly as the ballpark proposal.[17]

Since then Missoula has been an exemplar of the indecision that fails to promote the best of urbanity, but also prevents the worst of catastrophes. Commerce has invaded the once sleepy western edge of town and created the North Reserve corridor that foments car traffic and saps vigor from downtown retailing. But then a Pioneer League owner and the Arizona Diamondbacks decided to grace Missoula with the longed-for professional minor league baseball team.[18] Not-in-my-backyard opposition drove the hopefully named Ospreys from the site by the Clark Fork to one contemplated ballpark site after another. Again in 1994, Missoulians approved an $8 million bond to buy open space on hillsides and riverfronts and for playing fields.[19] But then in 1998 they turned down a request of $300,000 for parks maintenance.[20]

By 2002 the Ospreys were in their fourth season of playing in the inadequate ballpark of the local American Legion team. They finally started to build a stadium at the downtown site on the Clark Fork. But they needed $1.2 million for the next stage of construction. The director of the Diamondbacks minor league system said: "We've been patient to this point, but our patience is being tested."[21]

But then on June 25, 2004, the Ospreys played their first game in the ballpark by the river where they belong, within sight of an osprey nest.[22] The bleachers are sunk into the bench that borders the Clark Fork and form a bowl that opens toward the cottonwoods at the bank of the river. It's a handsome ballpark though only 1,900 of the bleachers were finished. They were filled to overflowing on that festive night. The 2004 season was not a successful one for the Ospreys in wins and losses. But they did draw a record crowd on opening night and for the entire season in their new place.[23]

Commodification and Realization

The realization of the Ospreys' dream illustrates how commodification and realization (in the specific sense I am using the word here) contrast with each other. The early promoters of a ballpark by the river had a sense of the crucial difference. An advertisement in support of the bond issue that was to acquire land for the ballpark began with the headline: "There's a lot more to 'Take me out to the ball game' than this." "This" was the screen of *Video Strike-Out* where in the bottom of the eighth inning the home team was behind five to thirty-nine. "This" was baseball as a commodity, available anytime and anywhere at whatever level with whatever outcome. "More" is a ballpark that marks a place and establishes a focal point in the community, one that gathers the river, the raptor, the team, and the citizens into one celebration at definite times in a specific season. As the advertisement put it in a mix of insight and stereotype:

> It's the fun of a family outing together, the excitement of the crowd, the taste of a ballpark hotdog and the once-in-a-lifetime thrill of catching a homerun ball. It's the way you feel when the whole town is pulling together, the joy of winning, the pain of defeat.
>
> This November 5th, we have the opportunity to give those experiences to our kids and our kids' kids, by voting Yes for the Four Seasons Sports Park [as the project was called before the Ospreys alighted on the site].
>
> And what's the price of this gift to ourselves and future generations?
>
> About a buck a month, if you own a home worth $50,000. That won't even buy a video cartridge, let alone the game itself.[24]

As the utter failure of this plea and the disastrous outcome of the elections of November 5, 1991, showed, the discourse of realization competes poorly with the juggernaut of commodification; and there is, not surprisingly, a like imbalance in the kinds of conduct that correspond to these contrasting designs—the imbalance of celebration and consumption.

The contrast between commodification and realization is a reminder of how important it is to distinguish, for all the overlap, between economic and moral commodification. The Interstate Highway System commodifies space morally. It more or less detaches space from its moorings and makes particular spaces comfortably available. But unlike toll roads, the interstates don't pull the traversing of space into the market. You do not have to pay directly so much per mile of travel. Conversely, the attendance at an Ospreys game is economically commodified, but not morally. You have to pay between $5 and $10 for admission. But what you witness is well placed in its context of time, space, and community.

The confusions and tensions between commodification and realization flared up in the wake of the attack on the Twin Towers of the World Trade Center on September 11, 2001. A catastrophe is often thought to clear a space for a new beginning. An empty space was literally opening up at Ground Zero as truck after truck carried off the ruins of those skyscrapers, piece by piece.[25]

The conflict between commodification and realization went through several stages. It began as the collision of two kinds of recollection—restoration and reverence, basically rebuilding what was there before vs. never building there again to honor the memory of destruction and death. Restoration was represented by the Port Authority of New York and New Jersey, the owner of the real estate, and Larry Silverstein, who had a lease on the land. Reverence was advocated primarily by the families of the victims.

Governor Pataki, who held the lion's share of political power regarding the fate of Ground Zero, appointed the Lower Manhattan Development Corporation (LMDC), a body of business, civic, and cultural leaders, that was to guide the reconstruction of the Trade Center site. The ensuing power struggle between the Port Authority and the LMDC ended with a memorandum wherein, as Paul Goldberger observed, "[t]he single most important aspect of the rebuilding process—the Port Authority's insistence on rebuilding all of the lost office space, as well as the retail and hotel space—went unchallenged."[26] From then on, the machinery of commodification overshadowed, if it didn't suffocate, realization and celebration.

The LMDC received hundreds of unsolicited suggestions from designers skilled and inept, high-minded and wacky, famous and obscure. It finally hired the architectural firm of Beyer Blinder Belle to draw up plans for reconstruction. Ambiguities, confusions, and intrigues began right away to settle on the rebuilding of Ground Zero. It was unclear whether Beyer Blinder Belle was expected to provide an outline or a plan, whether the firm was given enough time, and whether it was dealt with fairly. The proposal was taken to be a detailed plan and found to be trite. Hence plans from other architects were invited for an exhibit of altogether six alternatives. The public reacted with profound disappointment. The six proposals looked too conventional and too much alike.

In response, the LMDC sponsored "The Innovative Design Study." It developed into a call to seven well-known architects or teams of architects to present bold and creative proposals. It turned into a competition that was won by Daniel Libeskind. Although he smothered his design with sentimental and hyperbolic language, it was a creditable attempt at realizing and grounding a place of recollection and renewal. The two outstanding features were, first, to expose, down to bedrock, the slurry wall that amidst all the destruction had kept the Hudson at bay and to leave the bedrock that had supported the towers uncovered and untouched; and second, to erect a tower—the Freedom Tower, Libeskind called it—1,776 significant feet tall, whose uppermost part thrust a spire into the air much as the Statue of Liberty, only a few hundred yards away, holds up her torch.

Step by step, however, the requirements of commerce obscured and obliterated the claims of culture. The slurry wall could not remain exposed entirely because it needed lateral support. Bedrock could not be left uncovered because a subterranean space for transportation, parking, and utilities was needed. At first it seemed that the upper half of the wall could remain visible. The Innovative Design Study foresaw a separate competition for the memorial of 9/11. The memorial was the remainder of "vision" in its struggle with "the program" of the Port Authority, of the symbolic in its collision with the commercial, of architecture in its contest with money.[27] Ironically, the prevailing design obliterated the slurry wall entirely. It filled up Ground Zero to grade though it left the footprints of the towers visible as a pair of sunken reflecting ponds. Libeskind's Freedom Tower did not fare much better. David Childs, Larry Silverstein and the Port Authority's architect, took over its design and produced a skyscraper that was more practical and conventional though its torqued body and crowning web of cables were innovative. Libeskind's surging spire was reduced to a spike and has been imperiled by safety concerns.[28]

The embarrassment that overcomes us when we have to face the challenge of design happened in Boston too and in ways strikingly similar to those that have plagued New York City. Boston was given a blank slate when the Central Artery that had divided and defaced the city was replaced by the Big Dig. At twenty-five acres, the area, named the Rose Kennedy Greenway, is larger by nine than Ground Zero. In Boston, Mayor Thomas Menino should have provided leadership, but was slow in doing so. The idea of integrating the Greenway into the urban fabric was warmly welcomed. However what is needed in Robert Campbell's words "is a commanding idea that everyone can get behind. A civic vision."[29] Campbell applauded Ken Greenberg's proposal to make each of the crossings of the Greenway with major streets "a node of energy. Activity should concentrate there, and the energy would radiate along the cross street to tie the Greenway to other parts of the city."[30] But exactly what kind of centers the energy would emanate from is not clear. Meanwhile struggle for control and turf battles by adjoining neighborhoods are littering the Greenway.[31]

Centers of Communal Celebration

Public space, no less than the domestic sphere, is distorted and in disarray. That Ground Zero was a clearing for high-minded renewal and a challenge to set new standards was evident to many. Daniel Libeskind meant his design "to journey down into Ground Zero, onto the bedrock foundation" and to testify to "Life victorious" in the Freedom Tower.[32] Another of the celebrity architects, Lord Foster of Great Britain, proposed to create "sacred, cathedral-like spaces" under the gateway of his towers that were to connect and diverge in the shape of an X. As Goldberger saw it, "there was a genuine craving for an architectural response to the crisis, for creative designs that would somehow manage to demonstrate the ability of architectural aesthetics to heal a broken world."[33] And he thought that John C. Whitehead, chair of the LMDC, understood that the task before him called for "making harmony between the sacred and the ordinary."[34]

Like Missoula, New York City failed its highest aspirations but avoided moral disaster as well. It came to a conclusion that, though honorable, remains beset with uncertainty and indecision. There is of course a scale and grandeur to everything in New York that defies comparison with a little town in the Rockies. A skyscraper in particular is a most emphatic way of marking a spot and in its way the most American. As Goldberger has it, "the skyscraper is the ultimate American building type, the most important contribution of American architecture and American technology to world building."[35] All the major proposals for Ground Zero searched for novel skyscraper shapes. The problem with such innovative forms is that they leave the triteness of the content unquestioned. The content is trite in structure—a three-dimensional matrix of cells, so-called offices—and unspectacular in function—thousands of containers for lawyers, accountants, consultants, and such, to do their work. Attempts were made to place gardens, parks, and cultural spaces in the upper reaches of some of the towers.[36] None of these ideas survived.

If a skyscraper is an impressive form with meager content, the memorial park in Michael Arad's design is heavy with intent, but slight in form. Its intent is to recall the most violent attack on this country in nearly two centuries and to honor the victims of that violence. The form that the intent has been given is a little park with two square, depressed reflecting pools. The diffident form provides for little engagement. Visitors will walk around, look at the footprints, and remember 9/11. That's a good thing in a hyperactive and oblivious city, but it is less than a powerful response to an enormous event.

In the past, one reaction to this crisis of tradition and confidence has been an insistence on a massive rational and functional transformation of space. The clearing of a superblock for the geometrically austere and physically overpowering Twin Towers was itself a paradigm of this spirit. But the sterility of modernist designs turned out to be harmful and sometimes lethal to urban vitality. The World Trade Center was one of the last monuments of modernism. In time the brutality of its design was softened by habituation. We learn to come to terms with arrogance as long as there is enough grace in the surrounding city.

Another response to the spirit of uncertainty has been an unremitting rejection of tradition and of its aspiration to beauty or solemnity and a flaunting of cerebral severity that often borders on ugliness.[37] Little of this rebellion was found in the proposals for Ground Zero. The grief of the survivors assured a modicum of comity. A lot of this austere cerebration is academic. But it burst upon reality at least once when Richard Serra's *Tilted Arc* disfigured and disrupted a plaza in lower Manhattan.[38]

Similarly harsh reactions have occasionally been inflicted on private residences. But amidst all the domestic confusions and uncertainties, the dinner table has remained as a focal point of rest and celebration, if not always in practice, then at least in a persistence of aspiration. What's the public counterpart of the dinner table? In the everyday life of the city it is the fabric of small blocks and diverse uses, the fruitful messiness whose patron saint has been Jane Jacobs.[39] To repair the texture that had been intentionally erased by the designers of the World Trade Center and disastrously destroyed by the terrorists of 9/11 was a task that was honored by almost all proposals.

The good life in a good society needs more, however, than a lively urban fabric and the daily pleasures of walking, shopping, of seeing and being seen. There is a need for centers of celebration, for the public equivalents of the dinner table. Such centers of communal celebration are the places that are devoted to the arts, to athletics, and to religion. New York City is blessed with many such places. Central Park and the museums, concert halls, churches, synagogues, and mosques that surround it make Manhattan the magnificent city that it is. Of the three focal areas of celebration, Arad's memorial park is surely a creditable instance. It preserves and honors sacred ground and goes as far toward providing an inclusive place of reflection and reverence as is possible.

The New York City Opera would have been a more full-blooded cultural institution on Ground Zero. But it was thought to be physically and artistically too imposing. A consortium of smaller ventures was invited instead, a small dance company, a small theater company, and a small exhibition space.[40] It's one more instance of how we shape public space—we fail to reach for the best while avoiding the worst.

CHAPTER FIFTEEN

Realizing American Ethics

Reaching for Reality and Ethics

To discover the common and compelling vision of the good life and the good society that has beckoned and eluded us, we need to realize what ethics demands of us, in both senses of *realizing*—we have yet to recognize the need for it, and unsurprisingly, we still have to give it a commanding and tangible shape. We have lost touch with reality, a remarkable turn of events considering how compellingly real this continent used to be to its arrivals.

America was a new world when the first Americans crossed over the Bering land bridge some twenty millennia ago, and it was a new world for the European settlers who began to arrive on the east coast of its northern part four centuries ago. You don't take a new world for granted. It commands your attention on pain of starvation and death.

To survive you had to become intimate with American reality. The first Americans have set the standard for all to come. Their intimacy was such that they could flourish on this continent without having to deform it. Whether the Indians were good stewards of the land, whether they eradicated the megafauna they encountered on their arrival, to what extent their hunting, gathering, and agricultural practices transformed the flora and fauna of this land—these are controversial issues.

In any event, their traces on this continent were light compared with the imprint settlements left on Europe. Thus, when first arriving here, the Europeans thought they had come upon primeval nature. They in turn had to involve themselves deeply and often painfully with the reality of this continent, and it was their encounter with the Native Americans and their involvement with a new reality that made them realize they were Americans and not just, or rather than, Britons, Germans, or Scandinavians.

The word *real* has many meanings. In its widest sense, everything that can be thought, perceived, or felt is real. If it were not, it would be nothing to us. But *real* and *reality* also convey the more particular meaning of presence and validity. In that sense we distinguish real gold from fool's gold, admonish an evasive or deceptive person to get real, invoke the reality principle to call attention to inescapable trials and forces, and, when they have hit us, acknowledge a reality check.

In that particular sense we can say that American ethics was once distinguished by its intimacy with reality and is now threatened by a sense of unreality. This turn of events began with the conquest of nature that got underway just as the Europeans discovered America. The supposedly untamed American continent seemed manifestly destined for conquest and in the nineteenth century provoked gigantic efforts of industry and technology. In time the machinery of control became so sophisticated and extensive that by now it has largely overlain the original reality of nature and history. Today what we are typically in touch with are the machineries of production and the commodities of consumption.

Just as the decline of courage was noticed and mourned early in the modern period and forever after, so the waning of reality as commanding and engaging presence has been documented and was deplored already in post–Civil War America.[1] In fact the anxiety about losing touch with reality and the vigor it inspires was much more public and pronounced at the turn from the nineteenth to the twentieth century than it is today.[2] At times the commodification of reality is so subversive and complete today that actual reality seems to have slipped irretrievably from our grasp. Consider the "reality" television shows. They promise to put us back in touch with reality. But the commodifying eye of the television camera turns every reality into a commodity.

What's left are bursts of hunger for reality.[3] Some of us expose ourselves to the fury of white water or the pull of steep slopes. Icy wastes are sought out for their rage and severity. If you are close to the communities of extreme kayakers, skiers, or climbers, you know of more than a few who have paid for the challenge of reality with their lives.

The daring of these adventurers, though difficult to justify, is commonly matched and perhaps redeemed by incredible skills and enormous self-discipline. There are cheaper and clearly objectionable ways of getting to feel a forceful presence. One is taking drugs, another is gambling, a third is bad sex, a fourth is too much food—there must be fifty ways to leave your boredom. Although these ways of reaching for reality are destructive and deplorable, they have an air of plausibility, given the smother of daily ease and dullness.

There is a hunger for reality. There is also a hunger for ethics, or for "values" as we say these days, and it often is as restless and sometimes perverse as the hunger for reality. To have a cause, any cause, no matter how constricted, abstract, or counter to your best interests, is all that matters as long as there is friction, opposition, resentment, and the sense that you are up against something really big.

The detachment of ethics from reality is the popular counterpart to the distance between scholarly ethics and cultural reality. For most of us that gap does not open up as the craving for physical engagement or the zealous commitment to a cause. The common case looks more like moral malnutrition, a subclinical malaise that surfaces as uncertainty and unhappiness in the research of social psychologists. How do we cure this malady?

Reaching for Reform

The word *grace* is the focal point of many meanings. It can refer to the sparkle of a person and to the force that lends the person radiance. It can designate the spirit of a work of art and the excellence of an athlete, especially when shown under pressure. It can denote the charm of a doe and the prayer said before a meal. It can signify the presence of God and the practice of secular high-mindedness as it does when John Rawls concludes his great book with this remark on the standpoint of justice: "Purity of heart, if one could attain it, would be to see clearly and to act with grace and self-command from this point of view."[4]

Self-command may be within our power, but the power that can grace us is not. Hence grace as a virtue is readiness, but it is not certainty of success. There are many things we can do that make a graceful life impossible, but there is nothing we can do to guarantee a life or a society full of grace. That's what makes culture unpredictable and ungovernable.

It is important to keep this in mind when thinking about the improvement of the human condition. Grace as a blessing is both crucial for happiness and beyond the control of those who pursue happiness. To be so dependent grates against our expectations and demands. We may scorn Stalin's five-year and Hitler's four-year plans. But when it comes to proposals for reform, we are ruled by the ideal of the recipe—a list of ingredients and of instructions that, if conscientiously followed, will produce the result.

While we demand perfection of mechanics and procedures when it comes to social reforms, we lack all ambition as regards ends and substance. The items on political agendas are too narrow to make a moral difference to ordinary middle-class life even if they were perfectly enacted. Prayer in schools would make us no more religious. Bans on gay marriage do nothing to strengthen families. Outlawing abortion would not make us care more deeply about lives lost to disease and starvation.

Proposals of cultural reorientation face a particular challenge from the liberal side—the charge that they fail to advance "a program for social change."[5] In fact, however, programs of social reform are so easy to sketch. Here we go: Reinstate a substantial estate tax. Make the federal income tax more steeply progressive. Reinstate a vigorous capital gains tax. Reduce military spending. Use the revenue to institute universal health care patterned after the Canadian and British systems. Establish a minimum annual income. Build public housing. Rehabilitate the schools of the inner cities. Increase the pay of teachers in those schools. Raise the minimum wage and make the federal government the employer of last resort. Programs of environmental stewardship and international cooperation are just as easy to propose.

So what's the problem with these programs? They have no beginning, and they have no end. We don't know how to motivate people to support them, and we don't know what their moral purpose is should they ever be enacted. More ignorance, indifference, obesity, and passivity? The liberals who set these scruples aside and go to a full court press in support of social reform deserve our support. But so does reflection on whether it may not be the really good life that will furnish the source of strength and the final goal for social reform.

Moral success is not guaranteed, but neither is life entirely fickle fortune. We have learned, to be sure, that dinner can be a deadening ritual as the opening of *American Beauty* reminds us. A game can be boring, a hike painful, a performance tedious, worship dutiful. We can be sure, however, that if we never sit down to dinner, attend a game, or go on a hike, always shun concerts and avoid worship, that grace will rarely descend on us. If, to the contrary, we are faithful in our practices, good things happen more often than not. To make room for grace in our lives, then, is a sensible prescription, though not a fail-safe recipe. And to repeat a crucial point, we should not only ready our world for the focal things and communal celebrations that we know to have graced us in the past, we should also be ready to embrace innovation and creativity, novel events that we cannot predict but will know to be graceful when they come our way.

Both social justice and the good life need appropriate background institutions. Social justice, as Rawls and Roosevelt saw it, requires decent housing; effective systems of education and health care; the availability of good jobs; insurance against sickness, accidents, and unemployment; and security in old age.[6] The good life needs households centered on focal things and congenial to focal practices, and it needs a public sphere marked by centers of celebration and conducive to communal engagements.

Once the background institutions of social justice are in place, so is social justice itself, at least in the flexible and creative ways that Rawls envisioned under the heading of pure procedural justice. Yet when the background institutions of the good life have become reality, smugness can be the result. The graceful life needs the right institutions, to be sure, but it needs more—humility and a little skepticism, innovation and some daring, confidence and a bit of luck or, more correctly, of grace.

Contours of Real Ethics

There are signs, more subtle and hopeful than recklessness and zealotry, that a recognition of reality and ethics is emerging. The first of these surfaces in the notion of the sacred. The social world of religious liberals is divided along two crucial axes. One, obviously, divides society into theists or believers and unbelievers or atheists. The other, less overtly, divides society into those to whom certain things are sacred and those for whom some things are expensive and some not. There are three things that the former hold sacred: the welfare of every least and last human being, the integrity and beauty of nature, and the preservation of the great works of art.

A more sophisticated and fairer way of graphing the point would be a coordinate system with one axis going from atheism by way of agnosticism to theism, and another extending from devotion to commodification by way of mainstream decency to a regard for the sacred. Religious liberals would like to see themselves located as far out on the diagonal of the two axes as possible, looking with condescension on mercenary atheists. At any rate, the graph would permit infinitely varied and subtle social placements. But let me, for the sake of clarity, stick to the orthogonal dividing lines. Which of the two is more important, the one that separates me from George W. Bush or the one that separates me from Peter Singer? One way of answering the question is to return to the distinctions between legality, morality, and civility.

As to the first, legality is not an issue here. Both Bush and Singer are law-abiding citizens. Is the question one of private morality? Only by half. As regards our private convictions, Bush and I are professed Christians. As long as private morality is consistent with legality, it is not a matter for public reproach or condemnation. Moral convictions are certainly open to public discussion, disagreement, and attempts at gentle persuasion. But it is not right for Christians to denounce Muslims or atheists.

The question of what to hold sacred, to the contrary, arises in the realm of civility, whose standards of conduct, unlike laws, are neither compulsory nor, unlike personal convictions, optional. Regard for the sacred is a moral obligation, and we have a duty to reproach, if not also to condemn, those who are indifferent to misery and to grandeur. Reality engages us rightfully as the sacred.

Space is the primary dimension of the sacred, the space that envelops nature, works of art, and the human family entire. Contours of time also show a dawning recognition of reality. Here again an orthogonal graph reveals them best. The familiar line that divides political liberals from conservatives is thought to coincide with the division of the radical avant-garde from cultural traditionalists. The stereotypes are the political liberal who defends promiscuous sex and crucifixes immersed in urine and the political conservative who admires Michelangelo's Pietà and stands up for family values.

No doubt these two lines sometimes coincide. But they don't have to, and they shouldn't. Someone can be a political liberal who strongly supports an uncensored National Endowment for the Arts and gay marriage and at the same time is a cultural conservative who insists that Bach's *St. Matthew Passion* is the greatest work of art and that spouses, heterosexual and homosexual, should be faithful to one another. Cultural conservatism is the readiness to answer the claims and lessons of the history of this country and of one's particular tradition; it is the willingness to learn from the past; and it is the refusal to pass the liberal test of "How much cultural discomfort and aesthetic nausea can you bear with a smile?" Its cultural heroes are Wynton Marsalis and Mark O'Connor, artists of superb skill who can tell a revolution from revulsion.

The line between traditionalism and avant-gardism, too, lies in the region of civility. The law should not take sides, and traditionalists should always allow themselves to be surprised by avant-garde greatness.

The Dispersion of Real Ethics

The region of American society where a sense of real ethics is alive is the one where human welfare, nature, and art are held sacred and where the claims and lessons of history are honored. It cuts across the secular-religious and the liberal-conservative divisions and supports Morris Fiorina's point that this country is less divided than we think. Still, the conventional divides are so prominent as to obscure the boundaries of the region where real ethics may thrive. Hence it's hard to say how popular and populated that hopeful territory is. There are, however, well-defined and energetic movements within it.

Most prominent, perhaps, is the new urbanism. It is alive to the tradition of densely settled neighborhoods. It promotes actual contact with neighbors, and it invites residents to experience streets and plazas through walking. Just as significant is environmentalism in its loftier aspirations. It attempts to honor the dignity of nature and make people step out of their mechanized and mediated shells to get in touch with mountains and rivers. A third movement is staging a more general assault on the diversions and distractions that come between reality and us—the voluntary simplicity movement.[7]

All three of these movements have more specialized companions. The TV Turnoff Network "encourages children and adults to watch much less television in order to promote healthier lives and communities."[8] The Science and Environmental Health Network supports stronger connections between a healthy environment and a healthy life.[9] The slow food movement complies with the traditional ways of growing and cooking food, and it celebrates thoughtful dining.[10] The home power movement gets off the grid and devotes itself to producing and conserving energy on its own premises.[11] Juliet Schor's "downshifters" simplify their lives across the board to regain a calmer and more engaging life.[12]

In their awareness of real ethics, these movements are limited and scattered. They have their several platforms, memberships, and conventions. In one sense that's natural and inevitable. You can't do everything all at the same time. You need to focus, concentrate your expertise, and husband your resources. But all of these causes aim at the good life that is grounded in reality. The good life is a whole, however, or it is not good, and the kind of reality that can provide a restful place beyond the divisions into machineries and commodities must hang together.

Because these reform movements are dispersed, their effect on politics and society is weak. Just as the broad moderate middle is silent in politics and yields the public square to the extremists, so the dispersion of reform leaves the political agenda to commodification. Look at the presidential campaigns. The economy is always one of the top issues, and candidates fall all over themselves with their promises to make it grow and to raise prosperity. It's understood by everyone that this means more commodities and greater consumption. Recovering reality and curbing commodious consumption isn't part of any platform.

A Jeffersonian Life

Since the contours of real ethics are faint and the movements of reform dispersed, what's needed is concentration and illumination. A center is needed that collects the threads of inquiry and renewal and illuminates the wider context.

The dinner table is that focal thing, the center of grace where we can rest the case of our lives. When we gather around the table, we are where we want to be, with the people we love most, doing what we really enjoy. But granted all that, what is American about it? Don't people in hundreds of different countries do the same thing? Where is real *American* ethics?

The particular character of our ethics comes into focus through the American reality that is gathered in a household and at the table. We can think of that gathering as a Jeffersonian life. It can't be, to be sure, Jefferson's life. We cannot and don't want to live a life that literally resembles Jefferson's. We do not want to because we reject the slavery it rested on. We cannot because the cultural framework and the social conditions that need to be understood and reformed are so different from Jefferson's.

Learning from Jefferson is answering the claims and lessons of history; not using slavery, Jefferson's views of African Americans, and his relationship with Sally Hemings and their children as an alibi and a way to get out from under those demands and teachings is one of the things we have to learn. What Jefferson would want us to learn so we become "good and accomplished" as he put it in a letter to his daughter Martha, we can take from his instructions in that same letter:

> from 8. to 10 o'clock practise music.
> from 10. to 1. dance one day and draw another
> from 1. to 2. draw on the day you dance, and write a letter the next day.
> from 3. to 4. read French.
> from 4. to 5. exercise yourself in music.
> from 5. till bedtime read English, write &c.[13]

Evidently Jefferson wanted his daughter to be accomplished in music, to move with elegance, and to be well read and knowledgeable.

We can see the outlines of grace, courage, and wisdom in Jefferson's advice to Martha.

In one regard, this is very traditional instruction. Exactly 110 years earlier Gottfried Wilhelm Leibniz (1646–1716), Newton's rival in mathematics and Kant's predecessor in philosophy, set down similar advice for a prince, though Leibniz included fencing for the young gentleman's physical fitness and mathematics and the sciences in subjects of study.[14] (Martha, you have to know, was only eleven when she received Jefferson's letter while the prince was sixteen or seventeen.) Both Jefferson and Leibniz drew from an ideal of education and excellence that goes back to the portrait of the Renaissance person, drawn by Baldesar Castiglione in 1528.[15]

The lesson of history here is that what was once the privilege of the aristocracy is now an obligation of democracy—as a society we have the means and the duty to offer a princely education to all. To begin with knowledge and wisdom, Jefferson's own ideal of encompassing all significant knowledge is impossible now even as it was highly selective then. But some version of general education should be alive at the dinner table, and knowledge has to extend from there in widening circles.

For Americans the focal place of their lives is rarely defined by ancestry and inheritance. It's devotion to a place that makes you a resident. Few can claim an ancestral home from where to trace one's roots back over generations. But once settled, we should embrace the place with something of the dedication Jefferson showed in his *Notes on the State of Virginia* where he set down in affectionate detail what was known about the rocks, the water, the plants, the animals, the humans, and the culture of the place.[16]

The culture we are responsible for has a uniquely wide span in extent and variety. I recall a picture of late nineteenth-century Missoula, Montana. In the immediate foreground are railroad tracks. Behind them is the Gold Dust Hotel and Saloon and the Garden City Drug Store. Between the rails and the buildings, Flathead Indians with their travois are headed east across the Continental Divide to hunt the buffalo as they had done for thousands of years.[17] In this photograph, the ancient hunting and gathering culture intersects with the modern industrial culture. What in the Old World lies at a distance of millennia was alive within little more than a century on this continent. The ancestral conditions of the human family and the most advanced industrial culture lie close together in this country. In between are all the cultures that the several waves of immigration have carried to these shores.

All this constitutes contexts and centers for the culture of the table. Our most recent technological culture, due to its highly mediated and virtual character, brings the immediacy and actuality of the table and the meal, of family and friends into relief. The proximity of the ancient ground state of human culture reminds us of the presence reality once possessed and of the conditions we have evolved in to prosper—engagement with reality and closeness to our loved ones—conditions we can approach in the kitchen and the dining room.

The diversity of cultures returns us to the uniquely American task of choice and devotion and of memory and innovation. If you are of French descent, you can stick to French cuisine. But you can also draw on Asian cooking; and a person of Asian or diverse ancestry can embrace French cooking. The point is devotion to the craft and a practice of celebration.

The beginning of wisdom is to be broadly familiar with the width and depth of American culture and to realize deeply one of its possibilities at the dinner table. Whatever grace and friendship come from that realization, there needs to be the wider recognition of the economic and military power that has been generated by American culture, and there has to be an acceptance of the moral obligations that follow from power. Although the celebration of dinner should be wholehearted, it cannot be unreserved. Celebration has to imply the determination to widen the circle of well-being until it includes everyone in this country and on earth.

Fortitude has to go ahead of dining and follow it. The temptation of yielding to the comforts of fast food and relaxing entertainment is always there. If overcome through the practice of sitting down to dinner, fortitude is needed to be physically and publicly active afterward or at other times.

Once retired, Jefferson was in the happy situation of having made the exercise of fortitude that is such a challenge for us a normal part of his domestic life. "From breakfast, or noon at latest, to dinner," he wrote to Dr. Benjamin Rush in 1811, "I am mostly on horseback, attending to my farm or other concerns, which I find healthy to my body, mind and affairs."[18]

We, of course, have neither the estate nor the horses that Jefferson could rely on. Devoted practice must make up for less fortunate circumstances. We need to develop the habits of walking the neighborhood and the town as the new urbanism encourages us to do and to play in the parks, attend the concerts, visit the museums, and gather in the sanctuaries as vigorous urbanity enjoins us to.

But we have to venture further from our homes into the world of politics, and Jefferson painfully felt the tension between the household and "the theatre of public life."[19] Having served as minister to France for two years, he wrote to George Gilmer in 1787: "I am as happy no where else and in no other society, and all my wishes end, where I hope my days will end, at Monticello. Too many scenes of happiness mingle themselves with all the recollections of my native woods and fields, to suffer them to be supplanted in my affection by any other."[20] Jefferson's hope of return was disappointed when his commission was extended for two more years. He returned to the United States in 1789 only to learn that he had been appointed secretary of state. He resigned in 1793 when he wrote to Angelica Schuyler Church: "I am going to Virginia. I have at length been able to fix that to the beginning of the new year. I am then to be liberated from the hated occupations of politics, and to sink into the bosom of my family, my farm and my books. I have my house to build, my fields to form, and to watch for the happiness of those who labor for mine."[21] Yet a term as vice president and two terms as president were to follow, till finally in 1809 he was able to tell his neighbors of Albemarle County: "Long absent on duties which the history of a wonderful era made incumbent on those called to them, the pomp, the turmoil, the bustle and splendor of office, have drawn but deeper sighs for the tranquil and irresponsible occupations of private life, for the enjoyments of an affectionate intercourse with you, my neighbors and friends, and the endearments of family love, which nature has given us all, as the sweetener of every hour."[22]

Like Jefferson, we should center our lives in our homes, among family, friends, and neighbors. For us today, the ordinary obstacle is neither the struggle for survival nor the demands of public life, but the distractions of consumption. It takes fortitude to leave the cocoons of comfort and convenience. But once we have gathered at the dinner table, wisdom and friendship can be ours, and they in turn can give us the courage to join with our neighbors in the design of a public realm that encourages celebration. Perhaps we can draw from common celebrations the generosity and resourcefulness to meet our obligations of justice and stewardship. Thus the United States may become the country of grace that the people who came here have searched for and worked for.

Notes

Chapter One: Real American Ethics

1. Elaboration of the points made in the introduction and evidence and arguments to support them will be furnished in the subsequent chapters of the introduction and in parts 1 through 3.

Chapter Two: Decency and Passion

1. Avishai Margalit, *The Decent Society,* trans. Naomi Goldblum (Cambridge, MA: Harvard University Press, 1996), 1.

2. Margalit, *Decent Society,* 262.

3. On the inability of the right-wing conservatives to give their values legal reality, see Thomas Frank, *What's the Matter with Kansas?* (New York: Henry Holt, 2004), 121 and 208.

4. For the theorist's nightmare in astrophysics, see Steven Weinberg, *The First Three Minutes* (New York: Basic Books, 1993 [1977]), 88; Heinz R. Pagels, *Perfect Symmetry* (New York: Simon and Schuster, 1985), 264; Brian Greene, *The Elegant Universe* (New York: Norton, 1999), 168.

5. David S. Broder, "The Polarization Express," *Washington Post,* December 12, 2004, B7.

6. Morris Fiorina, *Culture Wars?* (New York: Pearson Longman, 2005), 7–21.

7. Fiorina, *Culture Wars,* 7–19, 34–52; Cass Sunstein, *The Second Bill of Rights* (New York: Basic Books, 2004), 63 and 269.

8. *Transparency International Corruption Perceptions Index 2004,* available at www.transparency.org, accessed January 15, 2005.

9. Fiorina, *Culture Wars,* 16.

10. John Dewey, *The Public and Its Problems* (Athens: Ohio University Press, 1999 [1927]), 113 and 114.

11. Dewey, *Public and Its Problems,* 116 and 126–27.

12. Ibid., 139.

13. G. A. Cohen, "Equality of What? On Welfare, Goods, and Capabilities," in *The Quality of Life,* ed. Martha Nussbaum and Amartya Sen, 9–29 (Oxford: Oxford University Press, 1993).

14. Amartya Sen, "Capability and Well-Being," in *Quality of Life,* ed. Nussbaum and Sen, 30–53, quote on 48.

15. Sunstein, *Second Bill of Rights,* 134–35.

16. Robert Putnam, *Bowling Alone* (New York: Simon and Schuster, 2000), 154–61.

17. The Endangered Species Act of 1973, available at http://endangered.fws.gov/esa.html, accessed December 22, 2005.

18. Gregg Easterbrook, "Here Comes the Sun," *New Yorker,* April 10, 1995, 38–43; Mark Sagoff, "Do We Consume Too Much?" *Atlantic Monthly,* June 1997, 80–96.

19. John Tierny, "Betting the Planet," *New York Times Magazine,* February 12, 1996, 52–53 and 74–81.

20. J. Baird Callicott, "Intrinsic Value in Nature: A Metaethical Analysis," *Electronic Journal of Analytic Philosophy* 3 (Spring 1995), 5, available at www.phil.indiana.edu/ejap/1995.spring/callicott.1995.spring.html, accessed July 5, 2000; Andrew Light and Eric Katz, "Introduction: Environmental Pragmatism and Environmental Ethics as Contested Terrain," in *Environmental Pragmatism,* ed. Light and Katz, 1–18 (London: Routledge, 1996), 2–3.

21. See Callicott, "Intrinsic Value in Nature," and also John O'Neill, "The Varieties of Intrinsic Value," *Monist* 75 (1992): 119–37.

22. Bill McKibben, *The End of Nature* (New York: Doubleday, 1990).

23. Paul W. Taylor, "The Ethics of Respect for Nature," in *Environmental Philosophy,* 2nd ed., ed. Michael E. Zimmerman, 71–86 (Upper Saddle River, NJ: Prentice Hall, 1990).

24. William Cronon, "The Trouble with Wilderness; or, Getting Back to the Wrong Nature," in *Environmental Ethics: Concepts, Policy, and Theory,* ed. Joseph DesJardins (Mountain View, CA: Mayfield, 1999), 371–82.

25. Cronon, "Trouble with Wilderness," 371, 372, 373, 378.

26. Cronon has recently moved away from constructivism and toward realism; see, e.g., his "Neither Barren Nor Remote," *New York Times,* February 28, 2001, A19.

27. David Strong, *Crazy Mountains: Learning from Wilderness to Weigh Technology* (Albany: State University of New York Press, 1995).

28. Caitlin E. Borgmann and Bonnie Scott Jones, "Legal Issues in the Provision of Medical Abortion," *American Journal of Obstetrics and Gynecology* 183 (2000): 584–94.

29. Some opponents of abortion distinguish between a human being and a person, and there can be endless hairsplitting in efforts to obscure the difference between a human zygote, embryo, or fetus and a human person.

30. Thomas of Aquino, *Summa Theologica,* part one, question 118, article 2; *Quaestiones Disputatae de Potentia Dei,* book one, question 3, article 9.

31. Fiorina, *Culture Wars,* 35.

32. Frank, *What's the Matter with Kansas,* 101.

Chapter Three: Kinds of Ethics

1. Robert Putnam, following an established convention, calls this social capital; see his *Bowling Alone* (New York: Simon and Schuster, 2000).

2. An alternative model is overlapping consensus; see John Rawls, "The Domain of the Political and Overlapping Consensus," *New York University Law Review* 64 (1989): 233–55.

3. Putnam, *Bowling Alone,* 290.

4. There is an exception to this in one version of one of the two dominant types of theoretical ethics, viz., utilitarianism. It is what Bernard Williams calls "Government House utilitarianism." It holds that ordinary people will act more reliably along utilitarian lines if they remain ignorant of the theory of utilitarianism; see his *Ethics and the Limits of Philosophy* (Cambridge, MA: Harvard University Press, 1985), 108–10.

5. Martin Heidegger, *Being and Time,* trans. John Macquarrie and Edward Robinson (New York: Harper and Row, 1962 [1927]); Michael Oakeshott, *On Human Conduct* (Oxford: Clarendon Press, 1975).

6. Charles Taylor, *Sources of the Self* (Cambridge, MA: Harvard University Press, 1989); Hubert L. Dreyfus, *Being-in-the-World: A Commentary on Heidegger's* Being and Time, *Division I* (Cambridge, MA: MIT Press, 1991).

7. See Rosalind Hursthouse, *On Virtue Ethics* (New York: Oxford University Press, 1999).

8. Aristotle, *Nicomachean Ethics,* 1904b, 13–26 in the Becker pagination.

Chapter Four: Moral Landmarks

1. Cass Sunstein, *The Second Bill of Rights* (New York: Basic Books, 2004), 161–62.

2. Sunstein, *Second Bill of Rights,* 12 and 182–83.

3. Immanuel Kant, *Grundlegung zur Metaphysik der Sitten,* ed. Karl Vorländer (Hamburg: Felix Meiner, 1962 [1785]), 29, my translation (vol. 4, 409, in the edition of the Prussian Academy [*AA*]). As for the necessity of a principle, see ibid., 3–7 (*AA,* 387–90); *The Foundations of the Metaphysics of Morals,* trans. Lewis White Beck (Indianapolis, IN: Bobbs-Merrill, 1959 [1785]), 26 and 3–6.

4. Kant, *Grundlegung,* 42 (*AA,* 421); *Foundations,* 39.

5. Kant, *Grundlegung,* 52 (*AA,* 429); *Foundations,* 47.

6. Kant, *Grundlegung,* 57 (*AA,* 434); *Foundations,* 52.

7. Kant, *Grundlegung,* 60 (*AA,* 436); *Foundations,* 54.

8. Kant, *Grundlegung,* 78–79 (*AA,* 452); *Foundations,* 71.

9. Kant, *Grundlegung,* 19–20, 55–56, 64–70 (*AA,* 402, 432–33, 440–46); *Foundations,* 18–19, 50–51, 59–64.

10. Walter G. Ong, *Orality and Literacy* (London: Methuen, 1982), 178–79; Lawrence Haworth, *Autonomy* (New Haven, CT: Yale University Press, 1986); Charles Taylor, *The Ethics of Authenticity* (Cambridge, MA: Harvard University Press, 1991).

11. This raises the specter of relativism or arbitrariness. Kant's solution is to claim that as rational moral agents we each give ourselves the same moral law. The compatibilist explication of autonomy rests on the diversity of possible social trajectories and respect for the contingency of the individual's situation—no one is in exactly your position and should forcibly restrict the way you work out, within the limits of legality, the meaning of your life.

12. Gordon G. Brittan Jr., "Autonomy and Authenticity," in *Wilderness and the Heart,* ed. Edward F. Mooney, 129–49 (Athens: University of Georgia Press, 1999), 131.

13. Paul Gordon Lauren, *The Evolution of International Human Rights* (Philadelphia: University of Pennsylvania Press, 1998), 205–40.

14. Lauren, *Evolution of International Human Rights,* 299–300.

15. Ibid., 220, 223, 225–26, 236–37.

16. Ibid., 283.

17. Ibid., 222.

18. Clare Murphy, "The Challenge of the 'Cannibal Consensus,'" available at http://news.bbc.co.uk/2/hi/europe/3437587.stm, accessed March 4, 2005.

19. Martha Nussbaum presents similar examples in "Human Functioning and Social Justice," *Political Theory* 20 (1992): 203–4.

20. Rosa Rojas, "Indigenous Autonomy in Chiapas: The Women Are Missing," available at www.eco.utexas.edu/faculty/Cleaver/bookend.html, accessed December 23, 2005.

21. R. G. Collingwood, *The Idea of History,* rev. ed. (Oxford: Clarendon Press, 1993 [1946]); Hans-Georg Gadamer, *Truth and Method* (New York: Crossroad, 1985 [1960]).

22. John Stuart Mill, *Utilitarianism* (Indianapolis, IN: Bobbs-Merrill, 1957 [1861]), 4–5; John Rawls, *A Theory of Justice,* 2nd ed. (Cambridge, MA: Harvard University Press, 1999 [1971]), 42–45.

Chapter Five: Jefferson and Kant

1. What about the reverse—was the United States foreign to Kant? Lewis White Beck has said that "Kant's enthusiasm for the French Revolution, the American Revolution, and the Irish effort to throw off the English yoke is well known"; see his "Kant and the Right of Revolution," in *Selected Essays on Kant,* ed. Hoke Robinson, vol. 6, North American Kant Society Studies in Philosophy (Rochester, NY: University of Rochester Press, 2002), 73-84, quote on 73. In Kant's published writings

there is a remark that shows Kant to have been familiar with the American Constitution; see Immanuel Kant, *Die Metaphysik der Sitten,* in *Werkausgabe,* ed. Wilhelm Weischedel, vol. 8 (Frankfurt: Suhrkamp, 1977 [1797]), 475 (vol. 6, 351, in the Prussian Academy edition [*AA*]). The published writings also show that Kant was conflicted about the right to revolution; see Beck, "Kant and the Right of Revolution."

2. Immanuel Kant, "Beantwortung der Frage: Was ist Aufklärung?" in *Werkausgabe,* ed. Wilhelm Weischedel, vol. 11 (Frankfurt: Suhrkamp, 1977 [1784]), 53–54, my translation (vol. 8, 35–37 in *AA*); "What Is Enlightenment?" in *Foundations of the Metaphysics of Morals,* trans. Lewis White Beck (Indianapolis, IN: Bobbs-Merrill, 1959 [1784]), 86–87.

3. Thomas Jefferson, "Response to the Citizens of Albemarle," in *The Portable Thomas Jefferson,* ed. Merrill D. Peterson, 259–60 (New York: Penguin, 1975 [1790]), quote on 259–60.

4. Immanuel Kant, *Grundlegung zur Metaphysik der Sitten,* ed. Karl Vorländer (Hamburg: Felix Meiner, 1962 [1785]), 22–23 (*AA,* vol. 4, 404); *Foundations,* 20. Kant credited Rousseau for having turned Kant from a scholar, contemptuous of the masses, to one who had learned "to honor human beings." *Bemerkungen in den "Beobachtungen über das Gefühl des Schönen und Erhabenen,"* ed. Marie Rischmüller (Hamburg: Felix Meiner, 1991 [1764–65]), 38; see also 130.

5. Jefferson to Peter Carr on August 10, 1787, *Portable Thomas Jefferson,* 424.

6. Kant, *Grundlegung,* 23 (*AA,* 404); *Foundations,* 21.

7. Jefferson to Carr, 425.

8. Jefferson to John Trumbull on February 15, 1789, *Portable Thomas Jefferson,* 435.

9. Jack McLaughlin, *Jefferson and Monticello* (New York: Henry Holt, 1988), 369–72.

10. McLaughlin, *Jefferson and Monticello,* 371.

11. Ibid., 371–72.

12. Kant, *Allgemeine Naturgeschichte und Theorie des Himmels, oder Versuch von der Verfassung und dem mechanischen Ursprunge des ganzen Weltgebäudes nach Newtonischen Grundsätzen abgehandelt,* in *Werke* (*AA*), ed. Prussian Academy, vol. 1, 215–368 (Berlin: Georg Reimer, 1910 [1755]).

13. Kant, "Allgemeine Naturgeschichte," 317.

14. Gordon G. Brittan Jr., *Kant's Theory of Science* (Princeton, NJ: Princeton University Press, 1978).

15. Alexander Pope, "EPITAPH: Intended for Sir Isaac Newton in Westminster-Abbey," *Selected Poetry,* ed. Pat Rogers (Oxford: Oxford University Press, 1996), 67.

16. Charles Donald O'Malley, *Andreas Vesalius of Brussels* (Berkeley: University of California Press, 1964), 150–81.

17. John Burd Carman, "Anatomist's Preface," in Andreas Vesalius, *On the Fabric of the Human Body,* trans. William Frank Richardson, vol. 1, xxxv–xli (San Francisco, CA: Norman, 1998).

18. Isaac Newton, *The* Principia, trans. I. Bernard Cohen and Anne Whitman (Berkeley: University of California Press, 1999 [1687]).

19. Ernest Nagel, *The Structure of Science* (New York: Harcourt, 1961), 153–74.

20. Kant, *Grundlegung,* 23–24 (*AA,* 405); *Foundations,* 21.

21. Kant, *Grundlegung,* 22 (*AA,* 404); *Foundations,* 20.

22. Kant, "Über den Gemeinspruch: Das mag in der Theorie richtig sein, taugt aber nicht für die Praxis," in *Kleinere Schriften zur Geschichtsphilosophie, Ethik und Politik,* ed. Karl Vorländer (Hamburg: Felix Meiner, 1913), 67–176 (*AA,* vol. 9, 273–313); "On the Proverb," in *Perpetual Peace and Other Essays,* trans. Ted Humphrey, 61–92 (Indianapolis, IN: Hackett, 1983).

23. Kant, "*Über den Gemeinspruch,*" 82–83 (*AA,* 286–87); "On the Proverb," 69–70.

24. Kant, "*Über den Gemeinspruch,*" 83 (*AA,* 87); "On the Proverb," 70.

25. Jefferson to Carr on August 19, 1785, *Portable Thomas Jefferson,* 381.

26. Jefferson to Carr on August 10, 1787, ibid., 424; see also ibid., 380–82.

27. Kant, *Grundlegung,* 22 (*AA,* 403–4); *Foundations,* 20.

28. Kant, *Grundlegung,* 60–61 (*AA,* 436–37); *Foundations,* 55.

29. Kant, *Grundlegung,* 43–46 (*AA,* 421–23); *Foundations,* 39–40.

30. Daniel C. Dennett, "The Moral First Aid Manual," in *The Tanner Lectures on Human Values,* vol. 8, 128–29 (Salt Lake City: University of Utah Press, 1988).

31. Arthur Schnitzler, *Leutnant Gustl,* part one available at http://gutenberg.spiegel.de/schnitzl/gustl/Druckversion_gustl.htm; part two available at http://gutenberg.spiegel.de/schnitzl/gustl/Druckversion_gustl2.htm, accessed December 23, 2005.

32. Gottfried Keller, "Der Landvogt von Greifensee," in *Sämtliche Werke,* vol. 2, 718–801 (Munich: Carl Hanser, 1958), 723.

33. Schnitzler, part two, 2 and 3.

Chapter Six: The Pursuit of Happiness

1. Thomas Jefferson to Benjamin Austin on January 9, 1816, in *The Portable Thomas Jefferson,* ed. Merrill D. Peterson (New York: Penguin, 1975), 547–50.

2. Henry Adams, *The Education of Henry Adams* (New York: Modern Library, 1996 [1905]), 11 and 61.

3. Adams, *Education of Henry Adams,* 32–34.

4. John Stuart Mill, *Utilitarianism,* ed. Oskar Piest (Indianapolis, IN: Bobbs-Merrill, 1957 [1881]).

5. Adams, *Education of Henry Adams,* 33, 72, 126, 192.

6. Ibid., 33 and 72.

7. Francis Fukuyama, "The End of History?" *The National Interest* 16 (Summer 1989): 3–18.

8. Mill, *Utilitarianism,* 10.

9. Garry Wills, *Inventing America: Jefferson's Declaration of Independence* (New York: Vintage Books), 240–55.

10. Jefferson to George Whyte on August 13, 1786, in *Portable Thomas Jefferson,* 399.

11. Mill, *Utilitarianism,* 10.

12. Ibid., 11.

13. John Stuart Mill, *On Liberty* (Indianapolis, IN: Bobbs-Merrill, 1956 [1859]).

14. John Rawls, *A Theory of Justice,* 2nd ed. (Cambridge, MA: Harvard University Press, 1999 [1971]), 19–22; Samuel Brittan, "Choice and Utility," in *Capitalism with a Human Face,* 65–84 (Hants: Edward Elgar, 1995), 65–68.

15. It has been controversial whether Americans on average have gained or lost leisure time in recent decades; see Juliet Schor, *The Overworked American: The Unexpected Decline of Leisure* (New York: Basic Books, 1992). John Robinson and Geoffrey Godbey found that we have gained leisure time: *Time for Life: The Surprising Ways Americans Use Their Time* (University Park: Pennsylvania State University Press, 1997). I have been persuaded by Robinson and Godbey.

16. Henry Sidgwick, *The Methods of Ethics,* 7th ed. (Indianapolis, IN: Hackett, 1981 [1907]), 131–50, 178, 477; Rawls, *Theory of Justice,* 78, 282–85; Brittan, "Choice and Utility," 68–84.

17. Bernard Williams, *Ethics and the Limits of Philosophy* (Cambridge, MA: Harvard University Press, 1985), 86.

18. Mill, *Utilitarianism,* 23.

19. Ibid., 25.

20. Adam Smith, *An Inquiry into the Nature and Causes of the Wealth of Nations* (New York: Modern Library, 1985 [1776]), 223. There is more to Smith's vision of the free market; see Amartya Sen, "Does Business Ethics Make Economic Sense?" *Business Ethics Quarterly* 3 (1993): 45–49.

21. Peter Singer, "The Singer Solution to World Poverty," *New York Times Magazine,* September 5, 1999, 63.

22. More precisely, the relevant population for Singer is the totality of animals that can feel pleasure or pain. It follows for a utilitarian that one should favor animals capable of much pleasure over humans destined to feel mostly pain.

23. Bernard M. S. van Praag and Paul Frijters, "The Measurement of Welfare and Well-Being: The Leyden Approach," in *Well-Being,* ed. Daniel Kahneman, Ed Diener, and Norbert Schwarz, 413–33 (New York: Russell Sage, 1999), 414–15; Bruno S. Frey and Alois Stutzer, *Happiness and Economics* (Princeton, NJ: Princeton University Press, 2002), 19–25.

24. John Dewey sketches this approach in *Human Nature and Conduct* (Carbondale: Southern Illinois University Press 1988 [1902]), 148–49.

25. Bernard Weintraub, "Hollywood Ending," *New York Times,* January 30, 2005, sec. 2, 15.

26. Robert E. Lane, *The Loss of Happiness in Market Democracies* (New Haven, CT: Yale University Press, 2000); David G. Myers, *The American Paradox* (New Haven, CT: Yale University Press, 2000); Gregg Easterbrook, *The Progress Paradox* (New York: Random House, 2003).

27. Williams, *Ethics and the Limits of Philosophy,* 81.

28. Ronald Bailey, "The Pursuit of Happiness," *Reason,* December 2000, 1, available at www.findarticles.com/p/articles/mi_m1568/is_7_32/ai_67589548/print, accessed December 29, 2005.

29. U.S. Census Bureau, *Statistical Abstract of the United States: 2003,* no. HS-33, available at www.census.gov/statab/hist/HS-33.pdf, accessed December 29, 2005.

30. "Changing Income Inequality in the United States," available at http://william-king.drexel.edu/top/prin/txt/factors/dist5.html, accessed July 14, 2004.

31. Klaus Deininger and Lyn Squire, "Measuring Income Inequality," Harvard Institute for International Development (Cambridge, MA: Harvard University), May 1996, Table 1. The data for the United States are from 1947 to 1991, for the United Kingdom from 1961 to 1991.

32. "A Quarter-Century of Growing Inequality," available at www.inequality.org/facts.html, accessed July 15, 2004.

33. Ibid.

34. There have been many attempts at devising a measure of national welfare that is more inclusive and culturally more telling than the standard of living (GDP over population); see Clifford Cobb, Ted Halstead, and Jonathan Rowe, "If the GDP Is Up, Why Is America Down?" *Atlantic Online,* October 1995, available at www.theatlantic.com/politics/ecbig/gdp.htm, accessed October 18, 2002; and Jason Venetoulis and Cliff Cobb, "The Genuine Progress Indicator 1950–2002 (2004 Update)," available at www.RedefiningProgress.org/newpubs/2004/gpi_march2004update.pdf, accessed March 14, 2005.

35. Michael Walzer, *Spheres of Justice* (New York: Basic Books, 1983), 100–103.

36. Richard Posner, "The Ethics and Economics of Enforcing Contracts of Surrogate Motherhood," *Journal of Contemporary Health Law and Policy* 5 (1989): 22. Posner goes on to argue that contracts must be enforceable to be mutually beneficial.

37. Aristotle, *Nicomachean Ethics,* trans. H. Rackham with my emendations (Cambridge, MA: Harvard University Press, 1962 [1934]), 1095a 20–30.

38. Daniel Kahneman, Alan B. Krueger, David A. Schkade, Norbert Schwarz, and Arthur A. Stone, "A Survey Method for Characterizing Daily Life Experience: The Day Reconstruction Method," *Science,* December 3, 2004, 1777.

39. F. Thomas Juster, "Preferences for Work and Leisure," in *Time, Goods, and Well-Being,* ed. Juster and Frank P. Stafford, 333–51 (Ann Arbor: Institute for Social Research of the University of Michigan, 1985), 335; Robinson and Godbey, *Time for Life,* 243.

40. Kahneman et al., "Survey Method," 1780.

41. Quoted by Benedict Carey in "What Makes People Happy? TV, Study Says," available at www.nytimes.com/2004/12/02/health/02cnd-mood.html?ei=5094$en=37a9ae3, accessed December 29, 2005.

42. See note 39 above.

43. Kahneman et al., "Survey Method," 1776 and 1777.

44. Quoted by Carey, "What Makes People Happy?"

45. Mihaly Csikszentmihalyi and Robert Kubey, *Television and the Quality of Life* (Hillsdale, NJ: Lawrence Erlbaum, 1990) and (by the same authors), "Television Addiction Is No Mere Metaphor," *Scientific American,* February 2002, 74–80.

46. Martin Seligman, *Authentic Happiness* (New York: Free Press, 2002), 81–82.

47. Aristotle, *Nicomachean Ethics,* 1100a 10–1101b 9.

48. Ibid., 1100b 8–11.

Chapter Seven: Evolutionary Psychology

1. David Brooks, "The Triumph of Hope over Self-Interest," available at www07.homepage.villanova.edu/satya.pattnayak/The%20Triumph%20of%20Hope%20Over%20Self-Interest.htm, accessed December 29, 2005; Thomas Frank, *What's the Matter with Kansas?* (New York: Henry Holt, 2004), 1–109.

2. Nelson W. Aldrich Jr., *Old Money* (New York: Allworth Press, 1996), 3–64.

3. George F. Will, *Statecraft as Soulcraft* (New York: Simon and Schuster, 1983), 118–21 and frequently in his columns.

4. David Brooks, *Bobos in Paradise: The New Upper Class and How They Got There* (New York: Simon and Schuster, 2000).

5. Herbert Croly, *The Promise of American Life* (Boston: Northeastern University Press, 1989 [1909]), 329, 428–41.

6. Edward O. Wilson, *Sociobiology* (Cambridge, MA: Harvard University Press, 1975).

7. Bertrand Russell, *Introduction to Mathematical Philosophy* (London: Routledge, 1993 [1919]), 71.

8. Robert Wright, *The Moral Animal* (New York: Random House, 1994).

9. Alan Sokal and Jean Bricmont, *Fashionable Nonsense* (New York: Picador, 1999).

10. Barbara Herrnstein Smith, "Is It Really a Computer? Presumption and Oversimplification in Steven Pinker's Description of the Mind," *Times Library Supplement,* February 20, 1998, 3–4.

11. Daniel C. Dennett, "The Moral First Aid Manuel," in *The Tanner Lectures on Human Values,* vol. 8, 119–47 (1988), and *Darwin's Dangerous Idea* (New York: Simon and Schuster, 1995), 494–510.

12. Dennett, *Darwin's Dangerous Idea,* 467–81.

13. Wright, *Moral Animal,* 102, 176, 257; Dennett, *Darwin's Dangerous Idea,* 461, 469, 470; David P. Barash, "Deflating the Myth of Monogamy," *Chronicle of Higher Education,* April 20, 2001, B17.

14. Charles Darwin, *The Descent of Man* (Amherst, MA: Prometheus Books, 1998 [1871]), 444–48.

15. Darwin, *Descent of Man,* 406.

16. For the structures that produce these colors, see Carol Kaesuk Yoon, "Scientists Uncover the Peacock's Most Colorful Secrets," available at www.nytimes.com/2003/10/28/science/earth/29PECO.html, accessed October 29, 2003.

17. Nicolaus Copernicus, "On the Revolutions of the Heavenly Spheres," in *Theories of the Universe,* ed. Milton K. Munitz, 149–173 (New York: Free Press, 1957), 150–51.

18. Richard Dawkins, *The Blind Watchmaker* (New York: Norton, 1996); see Michael Ruse, *The Evolution-Creation Struggle* (Cambridge, MA: Harvard University Press, 2005). Ruse presents a more circumspect and sophisticated version of the view I advocate; see also his *Can a Darwinian Be a Christian?* (Cambridge: Cambridge University Press, 2001).

19. Blaise Pascal, *Pensées,* ed. Victor Giraud (Paris: Rombaldi, 1935), 133 and 144–45 (nos. 358 and 416).

20. See note 8 above.

Chapter Eight: John Rawls

1. John Rawls, *A Theory of Justice,* 2nd ed. (Cambridge, MA: Harvard University Press, 1999 [1971]).

2. Rawls, *Theory of Justice,* 3.

3. Ibid., 53; final version on 266.

4. Ibid., 72; see also 53 and 266. Sometimes Rawls uses the term "difference principle" in a narrower sense where it designates the (a) part only.

5. Ibid., 242–51.

6. Cass R. Sunstein, *The Second Bill of Rights* (New York: Basic Books, 2004). As it happens, Rawls never mentions Roosevelt, Sunstein never mentions Rawls.

7. Mill and Rawls on one side and Kant on the other converge on this (liberal democratic) individualism from opposite positions—Mill and Rawls from the optimistic belief that the individual knows best and individualism will produce most happiness, Kant from the pessimistic conviction that the empirical world was too fickle a place to permit a rational pursuit of happiness so that it would be pointless, arrogant, and oppressive for society to tell individuals how to pursue happiness.

8. Rawls, *Theory of Justice,* 54–55 and 79.

9. Ibid., 374.

10. John Rawls, *Justice as Fairness,* ed. Erin Kelly (Cambridge, MA: Harvard University Press, 2001), 39 and 10.

11. Rawls, *Justice as Fairness,* 57.

Chapter Nine: Theory and Practice

1. Physicians are pressured by HMOs and by for-profit hospitals. Journalists are probably worse off; see Howard Gardner, Mihaly Csikszentmihalyi, and William Damon, *Good Work* (New York: Basic Books, 2001), 125–206; Russell Baker, "What Else Is New?" *New York Review of Books,* July 8, 2002, 4–8.

2. Immanuel Kant, *Grundlegung der Metaphysik der Sitten,* ed. Karl Vorländer (Hamburg: Felix Meiner, 1962 [1785]), 43–45 (vol. 4, 421–23, in the edition of the

Prussian Academy [*AA*]). *The Foundations of the Metaphysics of Morals,* trans. Lewis White Beck (Indianapolis, IN: Bobbs-Merrill, 1959), 39–41.

3. John Stuart Mill, *Utilitarianism* (Indianapolis, IN: Bobbs-Merrill, 1957 [1861]), 72 and 68–72.

4. Lawrence Lessig, *The Future of Ideas* (New York: Random House, 2001).

5. Deborah G. Johnson and Helen Nissenbaum, eds., *Computers, Ethics, and Social Values* (Englewood Cliffs, NJ: Prentice Hall, 1995).

6. Marie C. McCormick and Douglas K. Richardson, "Premature Infants Grow Up," *New England Journal of Medicine* 346 (January 17, 2002): 197–98.

7. Bill McKibben, *Enough* (New York: Henry Holt, 2003).

8. John Fitzpatrick, "Jodie and Mary: Whose Choice Was It Anyway?" in *Spiked Liberties,* available at www.spiked-online.com/Printable/00000000540E.htm, accessed December 30, 2005.

9. Peter Singer, *Practical Ethics,* 2nd ed. (Cambridge: Cambridge University Press, 1993); Hugh Follette, ed., *The Oxford Handbook of Practical Ethics* (Oxford: Oxford University Press, 2003).

10. Immanuel Kant, "Über den Gemeinspruch: Das mag in der Theorie richtig sein, taugt aber nicht für die Praxis," in *Kleinere Schriften zur Geschichtsphilosophie, Ethik und Politik,* ed. Karl Vorländer, 67–113 (Hamburg: Felix Meiner, 1913 [1793]), 82–83 (*AA,* vol. 8, 286–88); "On the Proverb," in *Perpetual Peace and Other Essays,* trans. Ted Humphrey, 61–92 (Indianapolis, IN: Hackett, 1983), 69–70.

11. Stephen Turner, *The Social Theory of Practices* (Chicago: University of Chicago Press, 1994).

12. Douglas Sloan, "The Teaching of Ethics in the American Undergraduate Curriculum, 1876–1976," in *Ethics Teaching in Higher Education,* ed. Daniel Callahan and Sissela Bok, 1–57 (New York: Plenum Press, 1980).

13. William DeWitt Hyde, *Practical Ethics* (New York: Henry Holt, 1921 [1892]).

14. Hyde, *Practical Ethics,* iii.

15. C. Wright Mills, *Sociology and Pragmatism,* ed. Irving Louis Horowitz (New York: Oxford University Press, 1966), 81–82.

16. Harold S. Kushner, *When Bad Things Happen to Good People* (New York: Avon, 1983 [1981]. On Randy Cohen, see Dan Seligman, "Moral Reckoning," *Commentary,* January 2003, available at www.commentarymagazine.com/pbk.seligman.htm, accessed January 6, 2003.

17. Robert B. Louden, "On Some Vices of Virtue Ethics," *American Philosophical Quarterly* 21 (1984): 227–36; Rosalind Hursthouse, *On Virtue Ethics* (Oxford: Oxford University Press, 1999); "The Grammar of Goodness: An Interview of Philippa Foot," *Harvard Review of Philosophy* 11 (2003): 32–44.

18. *General Education in a Free Society: Report of the Harvard Committee* (Cambridge, MA: Harvard University Press, 1945).

19. *General Education in a Free Society,* 81.

20. Ibid., 82.

21. Ibid., 89.

22. Jonathan Rose, "The Classics in the Slums," *City Journal,* Autumn 2004, available at www.city-journal.org/html/14_4_urbanities-classics.html, accessed December 30, 2005.

23. On this sea change, see Martin E. P. Seligman, *Authentic Happiness* (New York: Free Press, 2002), 125–29.

24. On the moral-nonmoral distinction, see William K. Frankena, *Ethics,* 2nd ed. (Englewood Cliffs, NJ: Prentice-Hall, 1973 [1963]), 62.

25. John Rawls, *A Theory of Justice,* 2nd ed. (Cambridge, MA: Harvard University Press, 1999 [1971]), 92.

26. Pico Iyer, "Summing Him Up," *New York Review of Books,* December 16, 2004, 75.

27. Martin E. P. Seligman, *Authentic Happiness* (New York: Free Press, 2002), 138.

Chapter Ten: Personal Virtues

1. William J. Bennett, ed., *The Book of Virtues* (New York: Simon and Schuster, 1993); André Comte-Sponville, *A Small Treatise on the Great Virtues,* trans. Catherine Temerson (New York: Henry Holt, 2001 [1996]).

2. There is voluminous research on wisdom. Most of it is antiquarian or divorced from contemporary culture. There is, moreover, a confusion in terminological practice. "Wisdom" is occasionally the translation for *phronesis*. In such cases it is often qualified as "moral" or "practical wisdom," and *sophia* is then rendered as "intellectual wisdom." I will use "prudence" for *phronesis* and "wisdom" for *sophia*.

3. Plato, *The Republic,* book four (419–45 in the Stephanus pagination).

4. Aristotle, *Nicomachean Ethics,* books one and ten (1094a–1103a and 1172a–1181b in the Becker pagination).

5. Proverbs 8 : 22–23; *New Revised Standard Version of the Bible* (Glasgow: Collins, 1989), 603.

6. Proverbs 8 : 29–31; *New Revised Standard Version,* 604.

7. Thomas of Aquino, *Summa Theologica,* second part of the second part, question 45, article 2 (my translation).

8. Michele Lamont, *Money, Morals, and Manners: The Culture of the French and American Upper-Middle Class* (Chicago: University of Chicago Press, 1992).

9. T. S. Elliott, *Collected Poems, 1909–1935* (London: Faber and Faber, 1936), 157.

10. Gene Bellinger, Durval Castro, and Anthony Mills, "Data, Information, Knowledge, and Wisdom," available at http://www.systems-thinking.org/dikw/dikw.htm, accessed December 30, 2005.

11. See my "The Headaches and Pleasures of General Education," *Montana Professor* 13 (Spring 2003): 10–15.

12. *General Education in a Free Society: Report of the Harvard Committee* (Cambridge, MA: Harvard University Press, 1945), ix.

13. Debra Bradley Ruder, "Faculty of Arts and Sciences to Vote on Revisions to Core Program," *Harvard University Gazette,* May 15, 1997, available at http://www.hno.harvard.edu/gazette/1997/05.15/FacultyofArtsan.html, accessed December 30, 2005. List of Core Courses available at www.registrar.fas.harvard.edu/Courses/Core/index.html, accessed December 30, 2005.

14. Core Program available at http://ict.harvard.edu/~core/redbook_2002.html, accessed February 10, 2003.

15. List of Core Courses, see note 13 above.

16. Report available at www.fas.harvard.edu/curriculum-review/report.html, accessed December 30, 2005. For criticisms of Harvard's general education, see Thomas Bartlett, "What's Wrong with Harvard?" *Chronicle of Higher Education,* May 7, 2004; Ross Douthat, "The Truth about Harvard," *Atlantic Monthly,* March 2005, 95–99.

17. Rodger Doyle, "Can't Read, Can't Count," *Scientific American,* October 2001, 24; "U.S. Kids Lagging in Math, Science," Associated Press, April 4, 2001, available at www.nytimes.com/aponline/national/AP-Math-Science-Scores.htm, accessed April 4, 2001; Floyd Norris, "U.S. Students Fare Badly in International Survey of Math Skills," *New York Times,* December 7, 2004, available at www.nytimes.com/2004/12/07/national/07student.htm, accessed December 7, 2004; Karen W. Arenson, "Math and Science Tests Find 4th and 8th Graders in U.S. Still Lag Many Peers," *New York Times,* December 15, 2004, available at www.nytimes.com/2004/12/15/education/15math.html, accessed December 15, 2004.

18. "Science and Technology: Public Attitudes and Public Understanding," National Science Foundation, available at www.nsf.gov/sbe/srs/seind02/c7/c7s1.htm, accessed April 2, 2004.

19. Michael X. Delli Carpini and Scott Keeter, *What Americans Know about Politics and Why It Matters* (New Haven, CT: Yale University Press, 1996), 87.

20. Carpini and Keeter, *What Americans Know,* 88.

21. Ibid., 68–69.

22. Ibid., 70, 75, 80, and 84.

23. Aristotle, *Nicomachean Ethics,* trans. H. Rackham, Loeb Classical Library (Cambridge, MA: Harvard University Press, 1962), 555 (1169a 20–25 in the Becker pagination).

24. Ibid., 157 (1115a 30–35).

25. Bernard Knox, Introduction, in Homer's *Iliad,* trans. Robert Fagles, 3–64 (New York: Viking, 1990), 25–30.

26. Kelly Rogers, "Aristotle's Conception of Tò Kalón," *Ancient Philosophy* 13 (1993): 355 n. 1.

27. Victor Davis Hanson, *The Western Way of War: Infantry Battle in Classical Greece,* 2nd ed. (Berkeley: University of California Press, 2000).

28. Aristotle, *Nicomachean Ethics,* 163 (1116a 30–35).

29. Thomas of Aquino, *Summa Theologica,* second part of the second part, question 142.

30. William Ian Miller, *The Mystery of Courage* (Cambridge, MA: Harvard University Press, 2000), 263–65.

31. Karl Marx and Friedrich Engels, *The Communist Manifesto,* ed. Samuel H. Beer (New York: Appleton-Century-Crofts, 1955 [1848]), 11.

32. William James, "The Moral Equivalent of War," in *Essays on Faith and Morals,* ed. Ralph Barton Perry, 311–28 (Cleveland, OH: Meridian Books, 1962 [1910]), 323.

33. James, "Moral Equivalent of War," 323.

34. Roger Scruton, "The End of Courage: The Strange Death of an Old Virtue," *Philosophy News Service,* June 4, 1999.

35. Miller, *Mystery of Courage,* 283.

36. Frederick Jackson Turner, "The Significance of the Frontier in American History," in *The Frontier in American History,* 1–38 (New York: Henry Holt, 1959 [1893]), 1.

37. Turner, "Significance of the Frontier," 2.

38. Ibid., 37.

39. Aristotle, *Nicomachean Ethics,* 493, 495, and 503 (1160b 30–1161a 9, 1161a 22–25, 1162a 15–34).

40. Ibid., 471 and 473, 483–575 (1158a 2–11, 1150b 15–1172a 16).

41. Georg Friedrich Wilhelm Hegel, *Grundlinien der Philosophie des Rechts* (Frankfurt: Suhrkamp, 1996 [1821]), 311; *Philosophy of Right,* trans. T. M. Knox (London: Oxford University Press, 1967), 11–12. Martin E. P. Seligman, in *Authentic Happiness* (New York: Free Press, 2002), 149, says "that arranged marriages in traditional cultures do better than the romantic marriages of the West."

42. Dan Hurley, "Divorce Rate: It's Not as High as You Think," *New York Times online,* April 19, 2005, available at www.nytimes.com/2005/04/19/health/19divo.html, accessed April 26, 2005.

43. David G. Myers, *The American Paradox* (New Haven, CT: Yale University Press, 2000), 36–59.

44. Charles Taylor, *Sources of the Self* (Cambridge, MA: Harvard University Press, 1989), 72.

45. Peter Laslett, *The World We Have Lost,* 3rd ed. (New York: Charles Scribner, 1984), 53–80 and 153–81. Similar marital and economic conditions could be found in the alpine village of Törbel in Switzerland from 1700 until the early twentieth century; see Robert McC. Netting, *Balancing on an Alp: Ecological Change and Continuity in a Swiss Mountain Community* (Cambridge: Cambridge University Press, 1981), 136–40 and 169–85.

46. Aristotle, *On the Soul,* 412a and 431b.

47. Kent C. Bloomer and Charles W. Moore, *Body, Memory, and Architecture* (New Haven, CT: Yale University Press, 1977), 5.

48. In 2004 it was 69 percent. *Statistical Abstract of the United States,* no. 1216, available at www.census.gov/statab/www, accessed December 30, 2005.

49. Bloomer and Moore, *Body, Memory, and Architecture,* 1–5.

50. Ibid., 1 and 3.

51. John Mack Faragher, "Bungalow and Ranch House," *Western Historical Quarterly* 32 (2001): 150, 153–55.

52. Faragher, "Bungalow," 165, 168, 171–72.

53. Julie V. Iovine, "The Last Gasp of the American Living Room," *New York Times on the Web,* January 28, 1999, available at http://www.nytimes.com/library/style/012899design-notebook.html, accessed December 30, 2005.

54. It was 98.3 percent in 1996; see *Statistical Abstract,* no. 915. Jane E. Brody "TV's Toll on Young Minds and Bodies," *New York Times on the Web,* August 3, 2004, available at www.nytimes.com/2004/08/03/health/03brod.html, accessed December 30, 2005.

55. John P. Robinson and Geoffrey Godbey, *Time for Life* (University Park: Pennsylvania State University Press, 1997), 136–53; Robert D. Putnam, *Bowling Alone* (New York: Simon and Schuster, 2000), 216–46.

56. Robinson and Godbey, *Time for Life,* 142–44.

57. "Digital Television," Federal Communications Commission, available at www.fcc.gov/cgb/consumerfacts/digitaltv.html, accessed December 30, 2005.

58. Eric A. Taub, "Pact Lifts an Obstacle to HDTV Transition," *New York Times on the Web,* January 2, 2003, available at www.nytimes.com/2003/01/02/technology/circuits/02teev.html, accessed December 30, 2005.

59. "Digital Television," Federal Communications Commission.

60. "About the FCC," available at http://hrauhfoss.fcc.gov/edocs/public/attachmatch/DOC-247863A1.pdf, accessed April 5, 2005.

61. Katie Hafner, "Let's All Gather Round the Screen," *New York Times on the Web,* February 5, 2004, available at www.nytimes.com/2004/02/05/technology/circuits/05thea.html, accessed December 30, 2005.

62. Mark Weiser, "Open House," available at www.ubiq.com/hypertext/weiser/wholehouse.doc, accessed April 5, 2005.

63. Mark Weiser and John Seely Brown, "The Coming Age of Calm Technology," available at www.ubiq.com/hypertext/weiser/acmfuture2endnote.htm, accessed December 30, 2005.

64. Weiser, "Open House."

65. Weiser and Brown, "The Coming Age."

66. Martyn Williams, "Move Into the Smart House of the Future," available at http://www.pcworld.com/resource/printable/article/0,aid,40381,00.asp, accessed August 1, 2001.

67. "Microsoft Tests PC-less Smart House," Associated Press story at http://www.usatoday.com/life/cyber/tech/review/crg976.htm, accessed August 17, 2001.

68. Bill Gates, *The Road Ahead* (New York: Viking, 1995), 218–19 and 221.

69. Ad van Berlo, "A 'Smart' Model House as Research and Demonstration Tool for Telematics Development," available at www.dinf.org/tide98/101/berlo_ad2.html, accessed August 17, 2001.

70. William J. Mitchell, *City of Bits* (Cambridge, MA: MIT Press, 1995), 44.

71. Lynn Neary, "Crass Elements Bound to Infiltrate your Children," National Public Radio, *Weekend Edition—Saturday,* November 1, 2003.

72. Mark Patinkin, "Much Ado about a Novel Idea," *Providence Journal,* December 12, 2002, G1.

73. Katie Hafner, "You There, at the Computer: Pay Attention," *New York Times on the Web,* February 10, 2005, available at www.nytimes.com/2005/02/10/technology/circuits/10info.html, accessed February 14, 2005.

74. Ibid.

75. Neary, "Crass Elements."

76. Scott Carlson, "Online Textbooks Fail to Make the Grade," *Chronicle of Higher Education,* February 11, 2005, A35–A36.

77. Jennifer Neils, *The Parthenon Frieze* (Cambridge: Cambridge University Press, 2001).

78. Bob Scriver, *Blackfeet: Artists of the Northern Plains* (Kansas City, MO: Lowell Press, 1990).

79. Claudia Roth Pierpont, "Jazzbo: Why We Still Listen to Gershwin," *New Yorker,* January 10, 2005, 77–78.

80. Jonathan Rose, "The Classics in the Slums," *City Journal,* Autumn 2004, available at www.city-journal.org/html/14_4_urbanities-classics.html, accessed December 30, 2005.

Chapter Eleven: Political Virtues

1. Genesis 6 : 9; Mark 6 : 20; Acts 10 : 22.

2. John Rawls, *A Theory of Justice,* 2nd ed. (Cambridge, MA: Harvard University Press, 1999).

3. So reported on the back cover of John Rawls, *Justice as Fairness,* ed. Erin Kelly (Cambridge, MA: Harvard University Press, 2001).

4. Robert Putnam, *Bowling Alone* (New York: Simon and Schuster, 2000).

5. Putnam, *Bowling Alone,* 54, 81, 84, 112, 124. There are many more instances of the second half of the curve.

6. Ibid., 247–84.

7. Ibid., 287–363.

8. Martin Gilens, "Political Ignorance and Collective Policy Preferences," *American Political Science Review* 95 (June 2001): 392 (mean: 26 percent of federal budget); Program on International Policy Attitudes, University of Maryland, "Americans on Foreign Aid and World Hunger," February 2, 2001, available at www.pipa.org/OnlineReports/BFW/Questionnaire.html, accessed April 6, 2005 (mean: 24 percent). The participants in the National Issue Convention, Philadelphia, PA, January 10–12, 2003, thought foreign aid was 1 percent of federal budget; see "Overview of Results," available at www.la.utexas.edu/research/delpol/bluebook%202003/Over2003.html, accessed April 6, 2005.

9. Gilens, "Political Ignorance," 386–87, and National Issues Convention, "Overview." Not so the people questioned by the Program on International Policy Attitudes.

10. Anup Shah, "Sustainable Development: The US and Foreign Aid Assistance," available at www.globalissues.org/TradeRelated/Debt/USAid.asp, accessed December 31, 2005.

11. Mark Danner, *The Massacre at El Mozote* (New York: Vintage Books, 1994). More atrocities that we are implicated in have meanwhile been committed, and Danner has provided another chronicle in *Torture and Truth: America, Abu Ghraib, and the War on Terror* (New York: New York Review Books, 2004).

12. Gregg Easterbrook, *The Progress Paradox* (New York: Random House, 2003), 41–45.

13. Putnam, *Bowling Alone,* 156.

14. Andrew Light and Eric Katz, "Introduction: Environmental Pragmatism and Environmental Ethics as Contested Terrain," in *Environmental Pragmatism,* ed. Light and Katz (London: Routledge, 1996), 1–18.

15. Bill McKibben, *The End of Nature* (New York: Doubleday, 1989).

16. Yale Center for Environmental Law and Policy, Yale University, and Center for International Earth Science Information Network, Columbia University, "2005 Environmental Sustainability Index," 26, available at www.yale.edu/esi, accessed December 31, 2005.

17. Eric Higgs, *Nature by Design* (Cambridge, MA: MIT Press, 2003), 214–23.

18. American Society of Civil Engineers, "2005 Report Card for America's Infrastructure," available at www.asce.org/reportcard/2005/index.cfm, accessed January 3, 2006; Katharine Barrett and Richard Greene in "The Government Performance Project" for 2005, available at www.governing.com/gpp/2005/intro.htm, accessed January 3, 2006, arrived at a grade of B- as an average for the fifty states.

19. Roy Want, "RFID: A Key to Automating Everything," *Scientific American,* January 2004, 56–65.

20. Bill Gates, *The Road Ahead* (New York: Viking, 1995), 157–83.

21. Want, "RFID: A Key to Automating Everything," 61.

22. Cass Sunstein, *Republic.com* (Princeton, NJ: Princeton University Press, 2002).

23. National Center for Health Statistics, "Health, United States, 2004," available at www.cdc.gov/nchs/hus.htm, accessed December 31, 2005.

24. Ali Mokdad, James S. Marks, Donna F. Stroup, and Julie L. Gerberding, "Actual Causes of Death in the United States, 2000," *JAMA* 291 (March 10, 2005): 1240.

25. Mokdad et al., "Actual Causes," 1238.

26. J. Michael McGinnis and William H. Foege, "The Immediate vs. the Important," *JAMA* 291 (March 10, 2005): 1263. The impact of overweight and obesity on health is controversial; see Katherine M. Flegal, Barry I. Graubard, David F. Williamson, and Mitchell H. Gail, "Excess Deaths Associated with Underweight,

Overweight, and Obesity," *JAMA* 293 (April 20, 2005): 1861–67, and Raja Mishra, "Study Is Seen as Clouding Risks to the Overweight," *Boston Globe online,* May 9, 2005, available at www.boston.com/yourlife/health/fitness/articles/2005/05/09/study_is_seen_as_clouding_risks_to_the_overweight/, accessed May 9, 2005.

27. Kelly D. Brownell and Katherine Battle Horgen, *Food Fight* (Chicago: Contemporary Books, 2004), 53–65.

28. P. Rozin, C. Fischler, S. Imada, A. Sarubin, and A. Wrzesniewski, "Attitudes to Food and the Role of Food in Life in the U.S.A., Japan, Flemish Belgium and France," *Appetite* 33 (1999): 168 and 175–76.

29. Mita Sanghavi Goel, Ellen P. McCarthy, Russell S. Phillips, and Christina C. Wee, "Obesity Among US Immigrant Subgroups by Duration of Residence," *JAMA* 292 (December 15, 2004): 2860–67.

30. Lizette Alvarez, "U.S. Eating Habits, and Europeans, Are Spreading Visibly," *New York Times on the Web,* available at www.nytimes.com/2003/10/31/international/europe/31OBES.html, accessed October 31, 2003; World Health Organization, "Controlling the Global Obesity Epidemic," available at www.who.int/nut/obs.htm, accessed June 3, 2004; see also note 27 above.

31. Rosin et al., "Attitudes to Food."

32. Ibid., 177.

33. Ibid., 166.

34. Germans, however, seem to behave more like Americans; see Bertrand Benoit, "Cheap and Cheerful," *Financial Times,* April 9/10, 2005, W1–W2.

35. Brownell and Horgen, *Food Fight,* 69–239.

36. Carl Hulse, "Vote in House Offers a Shield for Restaurants in Obesity Suits," *New York Times on the Web,* March 11, 2004, available at http://travel2nytimes.com/mem/travel/article-page.html, accessed April 15, 2005.

37. Christopher Lee, "Fast Food Chains Get a Break Today," *Washington Post,* March 11, 2004, A2.

38. Erica Goode, "The Gorge-Yourself Environment," *New York Times on the Web,* July 22, 2003, available at www.nytimes.com/2003/07/22/health/nutrition/22EATU.html, accessed July 23, 2003.

39. Brownell and Horgen, *Food Fight,* 207–10; David Barboza, "Rampant Obesity, a Debilitating Reality for the Urban Poor," *New York Times,* December 26, 2000, F5; Mary Duenwald, "Good Health Is Linked to Grocer," *New York Times,* November 12, 2002, F5.

40. Greg Critser, *Fat Land* (Boston: Houghton Mifflin, 2004), 7–19.

41. Critser, *Fat Land,* 10.

42. Barbara A. McCann and Reid Ewing, "Measuring the Health Effects of Sprawl," September 2003, available at www.smartgrowthamerica.org, accessed April 15, 2005; R. Sturm and A. Cohen, "Suburban Sprawl and Physical and Mental Health," *Public Health* 118 (2004): 488–96.

43. Robert Kirkman, "The Ethics of Metropolitan Growth: A Framework," *Philosophy and Geography* 7 (2004): 203.

44. Kirkman, "Ethics of Metropolitan Growth," 206.

45. Ibid., 209. For Kirkman's proposal of a framework to broaden and deepen the debate, see 210–17.

46. Andres Duany, Elizabeth Plater-Zyberk, and Jeff Speck, *Suburban Nation* (New York: North Point Press, 2000).

47. Brian E. Saelens, James F. Sallis, Jennifer B. Black, and Diana Chen, "Neighborhood-Based Differences in Physical Activity: An Environment Scale Evaluation," *American Journal of Public Health* 93 (2003): 1552–58.

48. James A. Levine, Lorraine Lanningham-Foster, Shelly K. McCrady, Alisa C. Krizan, Leslie R. Olson, Paul H. Kane, Michael D. Jensen, and Matthew M. Clark, "Interindividual Variation in Posture Allocation: Possible Role in Human Obesity," *Science* 307 (2005): 584–86.

49. For the concept of the technological fix see Alvin M. Weinberg, "Can Technology Replace Social Engineering?" in *Technology and the Future,* 6th ed., ed. Albert Teich, 30–39 (New York: St. Martin's Press, 1993).

50. Paul Goldberger, "It Takes a Village," *New Yorker,* March 27, 2000, 28–34; Sara Fairbrother, Karla Lubeck, Brian Pita, Meg Shelby, J. C. Solomon, "New Urbanism in Abacoa," available at www.cantanese.org/Abacoa/documents/New%20Urbanism%, accessed March 11, 2004; Bruce Podobuik, "The Social and Environmental Achievements of New Urbanism: Evidence from Orenco Station," available at www.lclark.edu/~podobnik/orenco02.bdp, accessed March 11, 2004; Jonathan Martin, "New Urbanism for Rural Communities," available at www.cdtoolbox.org/development_issues/000188.html, accessed April 18, 2005; Kevin M. Leyden, "Social Capital and the Built Environment: The Importance of Walkable Neighborhoods," *American Journal of Public Health* 93 (2003): 1546–51.

51. Martin, "New Urbanism," 2.

52. Michael Hodges, "Voter Participation Report," available at http://mwhodges.home.att.net/voting.htm, accessed April 18, 2005.

53. International Institute for Democracy and Electoral Assistance, "Voter Turnout from 1945 to Date," available at www.idea.int/vt/survey/voter_turnout_pop2.cfm, accessed December 31, 2005.

54. Michael Moss, "Both Parties See New Promise When the Ballot Is in the Mail," *New York Times,* August 22, 2004, 1 and 12.

55. Moss, "Both Parties See New Promise," 1.

56. Quoted by Jo Becker in "Voters May Have Their Say before Election Day," *Washington Post,* August 26, 2004, A1.

57. Voting Procedures Working Group, "The Constitution Project Reform Initiative," May 23, 2001, 1–2, available at www.constitutionproject.org/docs/Report_5-31.pdf, accessed April 18, 2005.

58. Curtis Gans in American Enterprise Institute, "Absentee Ballot Voting," October 19, 2004, available at www.dos.state.pa.us/election_reform/lib/election_reform/Absentee_Ballot_Voting_AEI_Panel_Transcript.pdf, accessed April 28, 2005.

59. Robert J. Sampson and Stephen W. Raudenbush, "Seeing Disorder: Neighborhood Stigma and the Social Construction of 'Broken Windows,'" *Social Psychology Quarterly* 67 (2004): 319–42.

60. Billie Giles-Corti and Robert J. Donovan have found that healthy walking is due in nearly equal parts to individual, social, and physical factors: "Relative Influences of Individual, Social Environmental, and Physical Environmental Correlates of Walking," *American Journal of Public Health* 93 (2003): 1583–89.

61. Daniela Deane, "Cherishing a Place at the Table," *Washington Post,* November 9, 2002, H1.

62. Matthew W. Gilman, Sheryl L. Rifas-Shiman, Lindsey Frazier, Helaine R. H. Rockett, Carlos A. Camargo, Alison E. Field, Catherine S. Berkey, and Graham A. Colditz, "Family Dinner and Diet Quality among Older Children and Adolescents," *Archives of Family Medicine* 9 (2000): 235–40; Marla E. Eisenberg, Rachel E. Olson, Dianne Neumark-Sztainer, Mary Story, and Lind H. Bearinger, "Correlations between Family Meals and Psychosocial Well-Being among Adolescents," *Archives of Pediatrics and Adolescent Medicine* 158 (2004): 792–96.

63. Eric Schlosser, *Fast Food Nation* (Boston: Houghton Mifflin, 2001), 4; Brownell and Horgen, *Food Fight,* 7–10 and 36–37; Critser, *Fat Land,* 180.

64. Amanda Hesser, "The Family That Eats Together . . . May Not Eat the Same Thing," *New York Times on the Web,* December 17, 2003, available at www.nytimes.com/2003/12/17/dining/17SERI.html, accessed December 17, 2003.

Chapter Twelve: Recognizing Reality

1. Andrew Feenberg, *Questioning Technology* (London: Routledge, 1999), 201–5.

2. John Rawls, *A Theory of Justice,* 2nd ed. (Cambridge, MA: Harvard University Press, 1999 [1971]), 15–19.

3. Jonathan Rose, "The Classics in the Slums," *City Journal,* Autumn 2004, available at http://www.city-journal.org/html/14_4_urbanities-classics.html, accessed January 11, 2005.

4. Laurel Thatcher Ulrich, *The Age of Homespun* (New York: Knopf, 2001).

5. Viviana A. Zelizer, "Circuits within Capitalism," available at www.sciences-sociales.ens.fr/textes/Zelizer.doc, accessed May 27, 2004.

6. Karl Marx, *Capital,* trans. Samuel Moore and Edward Aveling, ed. Friedrich Engels (London: Swan Sonnenschein, 1886), 1.

7. Karl Marx and Friedrich Engels, "Manifest der Kommunistischen Partei," *Werke,* vol. 4 (Berlin: Dietz, 1959), 469; Karl Marx, *Das Kapital, Werke,* vol. 23 (Berlin: Dietz, 1971), 86–87, 91–92.

8. Marx, "Ökonomisch-philosophische Manuskripte," *Werke,* vol. [40] (Berlin: Dietz, 1968), 510–12; Marx and Engels, "Manifest," 468–69; Marx, *Kapital,* 181–83.

9. Cass Sunstein, *The Second Bill of Rights* (New York: Basic Books, 2004), 54–55.

10. Marx, *Kapital,* 49–51.

11. Michael Walzer, *Spheres of Justice* (New York: Basic Books, 1983), 100–103.

12. David Noble, "Technology and the Commodification of Higher Education," *Monthly Review* 53 (March 2002): 26–40; "'Education Commoditization,' an Interview with David Noble," available at http://www.wildduckreview.com/interviews/noble.html, accessed September 15, 2003.

13. Lewis J. Perelman, *School's Out* (New York: William Morrow, 1992); Peter Drucker quoted by Robert Lenzner and Stephen Johnson in "Seeing Things as They Really Are," *Forbes*, March 10, 1997, 127.

14. Todd Oppenheimer, "The Digital Doctorate," *New York Times Education Life*, April 25, 2004, 31–33; Scott Jaschik, "Studying with Stanford-Columbia-Chicago," ibid., 33–34; Adam Liptak, "Forget Socrates," ibid., 34 and 41.

15. Fred Hirsch, *Social Limits to Growth* (Cambridge, MA: Harvard University Press, 1976), 71–114.

16. Mark A. Johnson, "Some Observations Concerning the Law of Surrogacy," at http://www.surrogacy.com/legals/article/checklist/chklst1.html, accessed January 3, 2006.

17. Margaret Jane Radin, *Contested Commodities* (Cambridge, MA: Harvard University Press, 1996), 135–44.

18. Elizabeth Anderson, *Value in Ethics and Economics* (Cambridge, MA: Harvard University Press, 1993), 170.

19. Richard A. Posner, "The Ethics and Economic of Enforcing Contracts of Surrogate Motherhood," *Journal of Contemporary Health Law and Policy* 5 (1989): 22.

20. Richard J. Arneson, "Commodification and Commercial Surrogacy," *Philosophy and Public Affairs* 21 (1992): 132–64.

21. Jesse S. Tatum, *Energy Possibilities* (Albany: State University of New York Press, 1995).

22. Merlin B. Brinkerhoff and Jeffrey C. Jacob, "Quality of Life and Alternative Lifestyle: The Smallholding Movement," *Social Indicators Research* 18 (1986): 153–73.

23. Radin so takes it on p. xiii.

24. On the different structures of allocation and distribution see Rawls, *Theory of Justice*, 241.

25. Zelizer argues that the discriminative theories rest on a Hostile Worlds distinction—mixing the discriminated realms contaminate both. The total theories she calls reductionist. For them, contrary evidence is "Nothing But" a special case of the one and only realm.

26. Radin, *Contested Commodities*, xii.

27. Ibid., 122. For a critique of an early version of Radin's theory, see Arneson, "Commodification and Commercial Surrogacy."

28. Anderson, *Value in Ethics and Economics*, 17; see also 141–67. For a critique of Anderson, see Adrian J. Walsh, "Teaching, Preaching, and Queaching about Commodities," *Southern Journal of Philosophy* 36 (1998): 433–52.

29. Arneson, "Commodification and Commercial Surrogacy," 139.

30. *Encyclopedia of Marxism*, available at http://www.marxists.org/glossary/about/index.htm, accessed May 13, 2004.

31. Anderson, *Value in Ethics and Economics,* 166–67, and Zelizer show that the boundary is blurred. Although this is an important point, it merely modifies the area of blameless modification.

32. Charles Taylor, *The Ethics of Authenticity* (Cambridge, MA: Harvard University Press, 1991), 6.

33. E. P. Thompson, "The Moral Economy of the English Crowd in the Eighteenth Century," *Past and Present* 50 (February 1971): 76–136.

34. Thompson, "Moral Economy," 83.

35. Ibid., 135.

36. Ibid., 83 and 98.

37. Ibid., 135.

38. Ibid., 43.

39. Ibid., 93.

40. Ibid., 45.

41. Ibid., 89–90.

42. Ibid., 47.

43. In an essay from which I have learned a lot, Paul B. Thompson broadly calls *structural* commodification what I call *economic* commodification, and *technological* commodification what I call *moral* commodification; see his "Commodification and Secondary Rationalization," forthcoming in *Philosophy of Technology: Building on Andrew Feenberg's Critical Theory of Technology,* ed. Tyler Veak (Albany: State University of New York Press).

44. Siegfried Giedion, *Mechanization Takes Command* (New York: Norton, 1969 [1948]).

45. Giedion, *Mechanization Takes Command,* 130–208.

46. Marx and Engels, "Manifest," 466.

47. Rebecca Mead, "Annals of Reproduction," *New Yorker,* August 9, 1999, 56–65; Larry Richter, "Tracking the Sale of a Kidney on a Path of Poverty and Hope," *New York Times,* May 23, 2004, 1 and 8.

48. "I'll keep 'em, thanks," *Missoulian,* January 8, 2002, B4.

49. Sherry Turkle, *Life on the Screen* (New York: Simon and Schuster, 1995), 9–73.

50. In this sense Hobbes speaks "of such things as are necessary for commodious living"; see Thomas Hobbes, *Leviathan,* ed. Kenneth Minogue (London: Everyman, 1994 [1651]), 73.

51. Martha Nussbaum, "Human Functioning and Social Justice," *Political Theory* 20 (1992): 203.

52. See, e.g., John Rawls, *Justice as Fairness,* ed. Erin Kelly (Cambridge, MA: Harvard University Press, 2001), 32–28 and 153–57.

53. John Stuart Mill, *Utilitarianism* (Indianapolis, IN: Bobbs-Merrill, 1957 [1881]).

54. Rosalind Hursthouse, *On Virtue Ethics* (Oxford: Oxford University Press, 1999).

55. Robert Putnam, *Bowling Alone* (New York: Simon and Schuster, 2000).

56. Lionel Trilling, "The Fate of Pleasure," in *The Moral Obligation to be Intelligent,* ed. Leon Wieseltier, 427–49 (New York: Farrar, Straus, Giroux, 2000 [1963]).

57. Martin Seligman, *Authentic Happiness* (New York: Free Press, 2002), 103.

58. The result is "eudaemonistic pessimism"; see Wilhelm Windelband, *A History of Philosophy,* vol. 1, trans. James H. Tufts (New York: Harper and Row, 1958 [1891]), 87.

59. Seligman, *Authentic Happiness,* 105.

60. Robert E. Lane, *The Loss of Happiness in Market Democracies* (New Haven, CT: Yale University Press, 2000); David G. Myers, *The American Paradox* (New Haven, CT: Yale University Press, 2000); Gregg Easterbrook, *The Progress Paradox* (New York: Random House, 2003).

61. Seligman, *Authentic Happiness,* 105.

Chapter Thirteen: The Economy of the Household

1. More on paradigmatic explanation and implication in my *Technology and the Character of Contemporary Life* (Chicago: University of Chicago Press, 1984), 68–78 and 101–7.

2. B. L. Rayner, *Sketches of the Life, Writings, and Opinions of Thomas Jefferson* (New York: Francis and Boardman, 1832), 524. This widely quoted remark may be apocryphal. Rayner mentions an unnamed and undated "visiter [*sic*]" as the source.

3. Eric Peter Nash, *Frank Lloyd Wright* (New York: Smithmark, 1996), 14.

4. On the financial woes, see Jack McLaughlin, *Jefferson and Monticello* (New York: Henry Holt, 1988), 377–79; Frank Lloyd Wright, *An Autobiography* (London: Longmans, Green, 1932), 110 and 114; John Lloyd Wright, *My Father Who Is on Earth* (Carbondale: Southern Illinois University Press, 1994 [1946]), 15.

5. Frank Lloyd Wright Home and Studio Foundation, *The Plan for Restoration and Adaptive Use of the Frank Lloyd Wright Home and Studio* (Chicago: University of Chicago Press, 1977), 9.

6. Thomas Jefferson, "Autobiography," in *The Life and Selected Writings of Thomas Jefferson,* ed. Adrienne Koch and William Peden, 3–104 (New York: Modern Library, 1998 [1944]), 51.

7. Frank Lloyd Wright, *An Autobiography,* 109.

8. Robert C. Twombley, *Frank Lloyd Wright* (New York: Harper and Row, 1973), 200.

9. Twombley, *Frank Lloyd Wright,* 43–44, 72–76, 202–3.

10. Robert Putnam, *Bowling Alone* (New York: Simon and Schuster, 2000), 98–102.

11. Neil Levine, *The Architecture of Frank Lloyd Wright* (Princeton NJ: Princeton University Press, 1996), 10.

12. Diane Maddex, *50 Favorite Rooms by Frank Lloyd Wright* (New York: Smithmark, 1998), 23, 43, 50, 53, 54, 59, 60, 67, 71, 73.

13. Martha Jefferson Randolph to Thomas Jefferson on January 31, 1801, and

Jefferson to Martha Jefferson Randolph on February 5, 1801, *The Family Letters of Thomas Jefferson* (Charlottesville: University of Virginia Press, 1986), 193 and 195.

14. John Lloyd Wright, *My Father,* 43.

15. Jefferson to John Adams on June 10, 1816, in *The Adams-Jefferson Letters,* ed. Lester J. Cappon (Chapel Hill: University of North Carolina Press, 1987 [1959]), 443.

16. *Thomas Jefferson's Library: A Catalogue with the Entries in His Own Order,* ed. James Gilreath and Douglas L. Wilson, Library of Congress, 1989, available at www.loc.gov.catdir/toc/becites/main/jefferson/88607928.toc.html, accessed October 14, 2004.

17. Virginia Randolph Trist, quoted in *A Day in the Life of Thomas Jefferson*: "A Delightful Recreation," available at www.monticello.org/jefferson/dayinlife/parlor/home.html, accessed January 3, 2006.

18. Jefferson in a letter to Giovanni Fabroni on June 8, 1778, in *The Portable Thomas Jefferson,* ed. Merrill D. Peterson (New York: Penguin, 1975 [1790]), 359, and in a letter to Nathaniel Burwell on March 14, 1818, in *Life and Selected Writings,* 629.

19. Helen Cripe, "Jefferson's Catalogue of 1783: Transcription of the Music Section," available at www.lib.virginia.edu/dmmc/Music/Cripe/cripe.html, accessed January 3, 2006.

20. Frank Lloyd Wright, *An Autobiography,* 21.

21. John Lloyd Wright, *My Father,* 21.

22. Brent Staples, "Lust across the Color Line and the Rise of the Black Elite," *New York Times,* April 10, 2005, section 4, 11.

23. Jefferson in a letter to Benjamin Austen on January 9, 1816, in *Portable Thomas Jefferson,* 547–50.

24. Jefferson's letter to James Oldham of January 1, 1805 is quoted and reproduced in McLaughlin, *Jefferson and Monticello,* 290–92.

25. Robert Campbell, "Why Don't the Rest of Us Like the Buildings the Architects Like?" *Bulletin of the American Academy* (Summer 2004): 22–25.

26. Jefferson to Adams on January 21, 1812, in *Adams-Jefferson Letters,* 291.

27. Staples, "Lust across the Color Line," 11.

28. Levine, *Architecture of Frank Lloyd Wright,* 24–25.

29. Frank Lloyd Wright, *An Autobiography,* 111–15. John Lloyd Wright, *My Father,* 25–52.

30. See Levine, *Architecture of Frank Lloyd Wright,* on Wright's "authenticity of place," 432–33.

31. John Lloyd Wright, *My Father,* 17.

Chapter Fourteen: The Design of Public Space

1. Robert Rhodes James, ed., *Winston S. Churchill: His Complete Speeches 1897–1963,* vol. 7 (New York: Chelsea House, 1974), 6869. Behind Churchill, there is John

Ruskin whose writings Churchill knew and appreciated; see Ruskin's *The Seven Lamps of Architecture,* 2nd ed. (New York: Dover, 1989 [1880]), where he speaks of "the distinctively political art of Architecture" (p. 2).

2. James, ed., *Winston S. Churchill,* 6869.

3. Ibid., 6870.

4. Ibid., 6871.

5. T. A. Heppenheimer, "The Rise of the Interstates," *Invention and Technology* (Fall 1991): 8–18.

6. This stretch of the Interstate was far from setting a record of costliness. The 12-mile section of I-70 through Glenwood Canyon in Colorado cost $40 million per mile. The 17.3-mile stretch of I-105 through Los Angeles from Norwalk to El Segundo cost $127 million per mile; see (no author indicated) "The Last, Best Interstate," *Compressed Air Magazine,* September 1993, 8–15, and James Anderson, "A Bitter Pill for Some, LA's Last Freeway Opens," *Missoulian,* October 14, 1993, C4.

7. Sherry Devlin, "No Stopping Now," *Missoulian,* September 8, 1991, E1 and E8.

8. Ginny Merriam, "City's Crystal Ball Not Very Good with Traffic," *Missoulian,* October 3, 2004, A1 and A9.

9. James, ed., *Winston S. Churchill,* 6870.

10. Guy Gugliotta, "With 'Scramjet,' NASA Shoots for Mach 10," *Washington Post,* November 10, 2004, A1.

11. Neal R. Peirce, "New Superhighways Threaten to Lead America to Dead End," *The [Raleigh, NC] News and Observer,* October 8, 1995, 29A.

12. Peirce, "New Superhighways," 29A.

13. Quoted in Richard Pérez-Peña, "I-287: Extend It and They Will Drive on It," *New York Times,* June 2, 1996, sec. 1, 35 and 40.

14. "First Needs, Then Wants," *Missoulian* editorial, October 31, 1991, A4.

15. Don Baty, "What Do We Owe?" *Missoulian,* October 15, 1991, B1 and B3.

16. Don Baty, "Missoula Voters Say No," *Missoulian,* November 6, 1991, B1.

17. "Voters Aren't Sold on Bond Issues," *Missoulian,* November 6, 1991, B1.

18. Kim Briggeman, "A Team of Our Own," *Missoulian,* November 17, 1998, A1 and A3.

19. Sherry Devlin, "Voters Say Yes to Open Space Bond," *Missoulian,* July 8, 1994, A1.

20. Bob Chaney, "Voters OK Aging Services Levy, Deny Parks," *Missoulian,* November 4, 1998, B1.

21. Kim Briggeman, "No Place to Nest," *Missoulian,* May 9, 2002, D3.

22. Rob Chaney, "New Stadium Opens for Missoula Osprey Baseball Team," *Missoulian News Online,* June 26, 2004, available at www.missoulian.com/articles/2004/06/26/news/top/news01.txt, accessed January 3, 2006.

23. Kim Briggeman, "O's Fail to Keep Up with Changing Times," *Missoulian News Online,* September 14, 2004, available at www.missoulian.com/articles/2004/09/14/sports/sports03/txt, accessed April 27, 2005.

24. *Missoulian,* November 3, 1991, D6.

25. In my discussion of Ground Zero I have drawn from Paul Goldberger, *Up from Zero* (New York: Random House, 2004) and Philip Nobel, *Sixteen Acres* (New York: Henry Holt, 2005).

26. Goldberger, *Up from Zero,* 89.

27. Ibid., 54, 87 89–90, 109, 235–48, 254–56; Nobel, *Sixteen Acres,* 46–47, 103, 150.

28. "Rethinking Ground Zero," *New York Times* editorial, April 24, 2005, sec. 4, 11.

29. Robert Campbell, "After the Big Dig, The Big Question," *Boston Globe,* May 26, 2002, available at http://nl.newsbank.com/nl-search/we/Archives, accessed November 18, 2004.

30. Robert Campbell, "His Design Turns the Artery into the Heart of the City," *Boston Globe,* March 25, 2004, D5.

31. Anthony Flint, "Parks Mired by Turf Battle," *Boston Globe,* June 13, 2004, available at www.boston.com/news/local/massachusetts/articles/2004/06/13/parks_plan_mired_by_turf_battle/, accessed January 3, 2006; "Greenway Weakness," *Boston Globe* editorial, April 4, 2005, available at www.boston.com/news/globe/editorial_opinion/editorials/articles/2005/04/04/greenway_weakness/, accessed January 3, 2006.

32. Goldberger, *Up from Zero,* 9; Nobel, *Sixteen Acres,* 138–39.

33. Goldberger, *Up from Zero,* 55.

34. Ibid., 64.

35. Ibid., 56; see also 198.

36. Nobel, *Sixteen Acres,* 137.

37. Nathan Glazer, "'Subverting the Context': Public Space and Public Design," *Public Interest* 109 (Fall 1992): 7–10.

38. Glazer, "'Subverting the Context,'" 10–15.

39. Jane Jacobs, *The Death and Life of Great American Cities* (New York: Random House, 1961).

40. Goldberger, *Up from Zero,* 244–47.

Chapter Fifteen: Realizing American Ethics

1. Alan Trachtenberg, *The Incorporation of America* (New York: Hill and Wang, 1982), 182–207; T. J. Jackson Lears, *No Place of Grace* (Chicago: University of Chicago Press, 1983), 103–39.

2. Lears, *No Place of Grace,* 103–7.

3. James Delingpole, "Anything to Escape the Tyranny of Comfort," *Sunday Times,* August 5, 2001, available at www.sunday-times.co.uk/news/pages/sti/2001/08/05/stirevnws01008.html, accessed August 10, 2001; A. Alvarez, "Ice Capades," *New York Review of Books,* August, 2001, 14–17; Betsy Cohen, "Engaged with Life, Feasting with Death," *Missoulian,* February 10, 2002, A1–A2.

4. John Rawls, *A Theory of Justice,* 2nd ed. (Cambridge, MA: Harvard University Press, 1999), 514.

5. Alan Wolfe, *Return to Greatness* (Princeton, NJ: Princeton University Press, 2005), 134.

6. Rawls, *Theory of Justice,* 242–51; Cass Sunstein, *The Second Bill of Rights* (New York: Basic Books, 2004), 13, 182–83.

7. Duane Elgin, *Voluntary Simplicity* (New York: William Morrow, 1981).

8. TV Turnoff Network, available at www.tvturnoff.org/index.htm, accessed January 28, 2005.

9. Science and Environmental Health Network, available at www.sehn.org/about.html, accessed January 28, 2005.

10. Carlo Petrini, *Slow Food* (New York: Columbia University Press, 2003).

11. Jesse Tatum, *Energy Possibilities* (Albany: State University of New York Press, 1995).

12. Juliet Schor, *The Overspent American* (New York: Basic Books, 1998), 111–42.

13. Letter to Martha Jefferson on November 28, 1783, in *The Portable Thomas Jefferson,* ed. Merrill D. Peterson (New York: Penguin, 1975), 366–67.

14. Gottfried Wilhelm Leibniz, *Allgemeiner politischer und historischer Briefwechsel,* ed. Akademie der Wissenschaften der DDR, vol. 1 (Berlin: Akademie-Verlag, 1986), 332–33.

15. Baldesar Castiglione, *The Book of the Courtier,* trans. George Bull (London: Penguin, 1976 [1528]). Jefferson apparently did not have a copy of this book in his library, but his catalogue had one entry (no. 60) titled "Instructions for a Young Nobleman" and another (no. 124) titled "Tracts in Ethics, to wit, Swedenburg [*sic*], Castiglione, Mably, Lites Forenses, etat [*sic*] primitif, slave trade, Benezet, etc. on Slavery." This part of his catalogue available at http://www.loc.gov/catdir/toc/becites/main/jefferson/88607928_ch16.html, accessed January 4, 2006.

16. Jefferson, "Notes on the State of Virginia [1781–82]," in *Portable Thomas Jefferson,* 23–232.

17. William E. Farr, *Folge der Spur des Büffels,* trans. Dietmar Kuegler (Wyk auf Foehr: Verlag für Amerikanistik, 2004), 52.

18. Jefferson to Dr. Benjamin Rush on January 16, 1811, in *The Life and Selected Writings of Thomas Jefferson,* ed. Adrienne Koch and William Peden (New York: Modern Library, 1998 [1944]), 556.

19. Jefferson, "To the Inhabitants of Albemarle County, in Virginia," April 3, 1809, in *Portable Thomas Jefferson,* 331.

20. Jefferson to George Gilmer on August 12, 1787, in *The Papers of Thomas Jefferson,* ed. Julian P. Boyd (Princeton, NJ: Princeton University Press, 1955), vol. 12, 26.

21. Jefferson to Angelica Schuyler Church on November 27, 1793, in *Papers* (1997), vol. 27, 449.

22. Jefferson, in *Portable Thomas Jefferson,* 330.

Index